FAQs About the Use of Animals In Science

A handbook for the scientifically perplexed

Ray Greek and Niall Shanks

Co-published by arrangement with
Americans for Medical Advancement

University Press of America,® Inc.
Lanham · Boulder · New York · Toronto · Plymouth, UK

University Press of America,® Inc.
4501 Forbes Boulevard
Suite 200
Lanham, Maryland 20706
UPA Acquisitions Department (301) 459-3366

Estover Road
Plymouth PL6 7PY
United Kingdom

Printed in the United States of America
British Library Cataloging in Publication Information Available

Library of Congress Control Number: 2009933215
ISBN-13: 978-0-7618-4849-3 (paperback : alk. paper)
ISBN-10: 0-7618-4849-5 (paperback : alk. paper)
eISBN-13: 978-0-7618-4850-9
eISBN-10: 0-7618-4850-9

Co-published by arrangement with
Americans for Medical Advancement

™ The paper used in this publication meets the minimum requirements of American National Standard for Information Sciences—Permanence of Paper for Printed Library Materials, ANSI Z39.48-1992

We dedicate this book to Jean and The Lummocks

Table of Contents

Preface

As this book goes to press, the United States finds itself in the midst of the most challenging economic crisis since the Great Depression. Few among us have been so fortunate as to be untouched by its many ramifications, whether it's a job layoff, diminished investment returns, or the prospect of an unmanageable home mortgage.

Perhaps one of the few benefits of these trying times is that they have sparked a new demand for greater transparency and accountability from our elected officials. Now more than ever, the American public wants to know why things are the way they are, how they got that way, and how they can change for the better. Meanwhile, debate about what is the best strategy for bringing about economic recovery and developing a more efficient government has shed a new and often unflattering light on how the U.S. Congress appropriates spending.

As more Americans examine how their tax dollars are being spent, new questions have been raised about where to invest government resources. Health care? Infrastructure? Education? Housing? Energy? Defense? Science and technology? All these issues pop up on the radar screen with predictable regularity—and rightly so, since they are crucial to the stability, security, and future of our country.

On one profound issue, however, few questions are ever raised, and that is the use of animals in science. As a result, the vast majority of Americans do not question whether or not billions of dollars of their tax dollars are being wasted on biomedical research modalities that have been shown to be scientifically invalid—specifically with regard to the use of animals as predictive models for human drug and disease response. While the ethics surrounding the use of animals in science does receive attention from the media and animal rights groups from time to time, the scientific basis for using animals in specific areas of science does not. Logically, one would want to know what areas of science can and do use animals successfully, and likewise what areas do not, before exploring the ethical implications of such uses. Nonetheless, such questions are usually skipped and the haranguing begins.

There are many scientifically valid uses for animals in science. We have no scientific quarrel with the use of animals in science where such uses are scientifically valid. We will leave the discussion of the ethics of such use to others while acknowledging that the suffering of animals does concern millions of people. However, the continued use of using animals as predictive models for human drug and disease response extracts a terrible cost, both in funds that could be used far more effectively elsewhere, and, more importantly, the harm and loss of human life it causes.

The simple reality is that this issue extends far beyond the mistaken perception that it is only for those who care about the treatment of animals. It is not an overstatement to say that it affects every single American living today. How ironic that this state of affairs goes largely unnoticed in an area that we all hold closest to our hearts, which is our health and well being. After all, what is more important to any of us than living a long, robust, and productive life?

It is a hallmark of our democracy that its citizens have a right—and we would argue a duty—to examine, understand, and evaluate programs that they are subsidizing through the taxes they are paying. With respect to the use of animals in biomedical research, however, there is no such demand for the transparency and accountability demanded in other areas. Meanwhile, researchers who continue to rely on the failed modality of animals as predictive models for human drug and disease response continue their work behind the closed doors of our most respected institutions in what can best be described as white coat welfare.

The other side of the coin is equally important. Many have rejected the use of all animals in science as scientifically invalid because they are morally disgusted by the practice. If they explained their position as being based on ethics, we would support their right to hold such an opinion. However, some claim that it is scientifically impossible to use animals in any area of study or education. While respecting their right to hold an opinion, we must differ with their conclusion.

One of the reasons why both sides in this debate are able to continue their diatribe is the fact that far too many people lack an interest in, let alone a fundamental understanding of, science. This is not entirely their fault.

Science education is neither mandatory nor even particularly emphasized in the U.S.—and most other countries, for that matter. If students do not learn about science, the scientific method, and critical thinking repeatedly throughout pre-college and college, they will probably complete their education with little actual understanding of science. As they move on through their lives, they're likely to be very intimidated by science and thus unwilling to read more on the subject or engage in scientific discussions. This lack of emphasis on science education plays well into the hands of any community with an agenda, because it limits open, educated discussion about the validity of whatever they oppose or promote. It also allows people who are antiscience to offer their misunderstandings about science to the society without being held accountable by society. The costs we pay here go well beyond a rational discussion of the uses of animals in science and include such matters as stem cell research and the development of reproductive and genetic technologies. In these cases, genuinely promising avenues of scientific research have been hampered by a prevailing climate of ignorance often cloaked in religious garb by ideologues who care for neither science nor the plight of people.

As a taxpayer, you are expected to foot part of the bill for scientific endeavors, and currently you are paying researchers' salaries without really knowing whether or not the money is being spent wisely. Now think of yourself as a consumer, where you evaluate claims of performance and value on a daily basis. Both as a taxpayer and a consumer, you have a right to know what you are buying.

If you had a company that manufactured a product that did not perform as advertised, you would eventually have to close your doors for lack of sales. That is essentially what we are saying about the use of animal models as predictive tools for human disease and drug response—that it does not work as advertised, and that it should be replaced with something that *does* work. This book is designed to help you make critical evaluations of the animal experimenters' claims, whose work is subsidized by your tax and charitable dollars.

We also examine the many ways in which science uses animals that are scientifically legitimate and attempt to address the underlying reasons why some have mistakenly denied that such use is scientifically viable. As this use of animals is less controversial, we will spend less time on it, but do not equate less space for less importance.

Before you can reasonably assess our claims you need to hear the other side of the arguments and we do our best to present those arguments or direct you to where those arguments are made. However, in order to make the appropriate evaluations, you need to understand what science is and how critical thought is used in science and in everyday life. In this book, you will learn the concepts of critical thought and science as they pertain to the use of animals in biomedical research. If you are one of the legions of individuals who are intimidated by science, you can rest assured that we have made every effort to present the material in as readable—and hopefully, interesting—a format as possible.

We have chosen a question-and-answer format precisely because it will help you move through the text more easily. You can read the book from front to back (which is what we suggest) or read certain sections that may be of particular interest to you. The Q&A format also makes it simple to use the book as a quick reference guide when you need to find an answer to a specific question.

Our critics may suggest that this book, which we have written for the layperson who wants to understand more about the use of animals in science, is too simplistic to be accurate. In one sense, they are right. To truly understand science and scientific concepts and facts, one really does need to be educated in biology, chemistry, physics, and critical thought at least at the level of first-year college course work in these subjects.

However, it is impractical to expect everyone to have this level of scientific knowledge. Our hope is that we will give you enough of an understanding of the issue, without overwhelming you with large amounts of complex data, to spark further reading on the topic and to gain enough confidence to engage in and contribute to conversations about animals, science, and public policy.

No one would expect you to hold your own in a formal debate with a PhD just because you have read this book. But if we can help you become a more enlightened, informed citizen, one who may even feel empowered to take a more active role in public policy matters, then we feel we have accomplished our goal.

(With regard to the reading level of this book, we might also add that if one is not able to explain one's life's work in a manner intelligible to an interested lay person, then one does not have a very strong grasp of what one is doing in the first place. A fundamental tenet of science and medicine is that if you can't explain it to a lay person, you don't understand it yourself.)

Some of you may be familiar with some of our previous publications, for example *Sacred Cows and Golden Geese* (Continuum 2000) and *Specious Science* (Continuum 2002), which one of us (RG) co-authored with his wife Jean Swingle Greek, DVM. And you may be wondering why there is any need for another book that covers essentially the same ground. The fact is that after *Sacred Cows* and *Specious Science* were published, the strongest arguments against using animals as models of humans began to come from the Human Genome Projects as well as advancements in such burgeoning fields as evolutionary and developmental biology, molecular biology, and genetics.

The arguments made in *Sacred Cows* were based primarily on clinical experience common to physicians or veterinarians. In fact, that book traces its roots to 1989, when I was completing my residency in anesthesiology and serving my first year as an anesthesiologist on staff at the University of Wisconsin while my wife, Jean, was completing her final year of veterinary school there.

During that time, we were both struck by how differently we treated the same diseases in animals and humans. Some medications that caused severe side effects in animals could be given to humans, and vice-versa. In light of these and other differences between humans and animals, we began to question how animal models could be used to cure human disease. As the year progressed, we began to take special note of the differences between humans and animals. For example, some tumors that are lethal in humans disappear without intervention in animals. Anesthetics that are commonly used in human patients must be avoided in some animal patients.

These observations led us to base *Sacred Cows* primarily on clinical experience common to physicians or veterinarians. In the practice of medicine, empirical data counts more than theory, as many treatments have looked good on paper (in theory) but failed miserably in practice. However, in most of the hard sciences (e.g., physics and chemistry) theory is more respected. At about the same time, one of the present authors (NS) was exploring the implications of the theory of evolution for the biomedical prediction problem (essentially the cluster of puzzles surrounding the use of animal subjects of one species to make predictions about likely test results in animals of another species with a different evolutionary history).

There are good scientific reasons for emphasizing the importance of our current best scientific theories. The Second Law of Thermodynamics, for example, predicts that a perpetual motion machine is impossible, which eliminates the need to test every new design for a perpetual motion machine. In the same way, the Theory of Evolution knits together the different branches of biology and allows predictions to be formed and tested.

In the following pages, we will present both empirical evidence *and* theory in explaining why, although there are many scientifically valid uses for animals in science, using the animal model as a method for predicting human drug and disease response is not one of them. Along the way, we use a variety of easy-to-read quotes from the scientific and lay literature—all with the proper context provided—as a way to help illuminate the points we are making. While the quotes may not be science *per se*, and no serious scientist would mistake them for scientific proof, they nonetheless can be quite revealing. This is particularly true in those instances where the scientific literature questions the value of the animal model in predicting human drug and disease response, thus refuting the argument that we are a lone voice in the wilderness.

In the end, we have attempted to present the topic in as understandable a way as possible for readers who are just beginning their exploration of the topic. For those who would like to go further, we direct you to *Animal Models in Light of Evolution* (Brown Walker 2009), written by the authors of the present book, which addresses many of the same issues from a more technical scientific perspective.

We hope that you find this book to be an adequate explanation of the issues surrounding the valid and invalid use of animals in biomedical research. And perhaps you will enjoy learning a little bit more about science along the way.

Ray Greek
Goleta, CA
April 30, 2009

Acknowledgements

The authors wish to thank Peggy Cunniff and Clare Haggarty for reading and commenting on this manuscript.

We thank Rita Vander Meulen for her assistance in preparing the manuscript and for her valuable comments and direction.

The diagrams of DNA, Gene, Chromosome and Gene Expression are reproduced here courtesy of the National Human Genome Research Institute, as indicated in most of those diagrams.

The table for NIH distribution of grants was provided by the National Academies Press, and we thank them for allowing us to reproduce it.

Portions of the sections on prediction came from "Are animal models predictive for humans?" (http://www.peh-med.com/content/4/1/2) published by Philosophy, Ethics, and Humanities in Medicine (http://www.peh-med.com/home/).

The authors gratefully acknowledge the National Anti-Vivisection Society of Chicago IL (www.navs.org) whose financial support helped make this book possible. The views expressed herein however, are solely those of the authors and do not necessarily reflect the views of the Society.

Chapter 1. An Overview of the Use of Animals in Science

How are animals used in science today?
Animals are used in science in at least nine different ways:

1. Animals are used as predictive models of humans for research into such diseases as cancer and AIDS.
2. Animals are used as predictive models of humans for testing drugs or other chemicals.
3. Animals are used as "spare parts", such as when a person receives an aortic valve from a pig.
4. Animals are used as bioreactors or factories, such as for the production of insulin or monoclonal antibodies, or to maintain the supply of a virus.
5. Animals and animal tissues are used to study basic physiological principles.
6. Animals are used in education to educate and train medical students and to teach basic principles of anatomy in high school biology classes.
7. Animals are used as a modality for ideas or as a heuristic device, which is a component of basic science research.
8. Animals are used in research designed to benefit other animals of the same species or breed.
9. Animals are used in research in order to gain knowledge for knowledge sake.

Are all of these scientifically valid?
It is the authors' position that the first two listed above *are not* scientifically valid uses of animals in science; however, the remaining seven *are* scientifically viable. We will discuss the scientifically viable uses of animals in science in greater detail in the next chapters. Later in the book, we will discuss why the use of animals as predictive models for human drug and disease response is not a scientifically valid modality.

Why is it important to delineate the different ways animals are used in science?
Failing to separate the distinct uses of animals in science can and often does result in fallacious reasoning when either defending the use of animals in science or arguing against it. The animal experimentation community, for

example, usually groups everything into one general category, then shows that one of numbers 3-9, say dissection, is scientifically viable. This creates the impression that the entire enterprise is viable—an example of the fallacy of equivocation. (You will learn more about the fallacy of equivocation in a later chapter. Suffice for now to say that the fallacy of equivocation occurs when a word of phrase is used differently in the premise than in the conclusion or subsequent premises.)

On the other side of the coin, less honest factions of the animal protection community often commit the fallacy of neglect of negative instances when arguing that all use of animals in science is scientifically invalid. They point out where animals have failed to predict or even correlate with human response but ignore the scientifically viable uses, because that does not favor their position.

There seems to be a lot of words and phrases used to describe the use of animals in science. Is there any difference between the words vivisection, animal models, animal-based studies, animal experimentation, and animal research?

Yes, there is, and it is important to learn the correct definition of each. Before we go any further in our discussion, we will define these terms in the ways they are used in everyday speech and how the authors will use them in this book. Later, you will see how the meanings of these words are often misconstrued not only by the animal experimentation community, but also by the animal protection community to prove their respective points. Our use of these words and phrases is based on the proper use and rules of English, and should be considered correct regardless of how others unintentionally misuse them or intentionally misconstrue them for spin purposes.

What is the definition of the phrase* animal research*?

When we use the phrase *animal research* we mean research involving animals when the research is meant to benefit the animal being used, or at least is not intended to harm the animal.

This use of the term is consistent with the phrase *human research*, which is an ethical practice that is conducted every day at hospitals and research institutions. Human research is the foundation of the field of epidemiology, which involves the study of the distribution of disease in human populations and the factors that influence the occurrence of disease, such as heredity, behavior, and exposure to environmental hazards.

Clinical studies are another part of human research. Clinical studies involve the observation and analysis of human patients. Since earliest times, they have provided the most accurate and usable clues to revealing the mysteries of the human body and the causes and effects of disease. In *clinical trials*, human volunteers, having provided informed consent, test out new treatments and medications under strictly controlled guidelines.

What is the definition of the phrase* animal experimentation*?

Animal experimentation is used to indicate research that is not necessarily meant to benefit the individual subject involved. Animals used in the context of animal experimentation are typically euthanized when the study comes to an end.

Human experimentation can be defined in much the same way, perhaps most appropriately by example. The experiments that German doctor Josef Mengele performed on concentration camp prisoners during World War II would be described as human experimentation. These prisoners were denied any choice in the matter, and the studies Mengele conducted were not designed to benefit them in any way.

Another example of human experimentation would be the *Tuskegee Study of Untreated Syphilis in the Negro Male*, conducted between 1932 and 1972 in Tuskegee, Alabama by the U.S. Public Health Service. In this study, researchers used almost 400 poor and mostly illiterate sharecroppers to study the natural progression of syphilis if left untreated. The men were neither required to give informed consent, nor were they even informed of their diagnosis. By 1947, when penicillin had become the standard treatment for syphilis, the researchers withheld penicillin (as well as any information about the treatment) from the study subjects so that they could continue studying how the disease spreads and kills.

What is the definition of the phrase* animal model*?

We use the phrase *animal model* to mean the use of an animal to model or represent humans. Go back to the first question in this chapter. Numbers 1 and 2 in the answer are an example of this use:

1. Animals are used as predictive models of humans for research into such diseases as cancer and AIDS.
2. Animals are used as predictive models of humans for testing drugs or other chemicals.

In numbers 3 through 9 of that same answer, animals would not be models of humans *per se*, at least causally—meaning that what causes a reaction in the animal will, by analogy cause the same reaction in a human. (Although in some cases, for example as heuristic devices, animals could still be considered models, but we will discuss that in more detail in the next chapters.)

What is the definition of the term* vivisection*?

Vivisection literally means dissecting or cutting up the living (as opposed to *dissection*, which is the cutting up of the once living). Technically, the word can and has been used correctly to describe the practice both in animals and humans. In the 1800s, the word vivisection was used by almost everyone, but today the term is mostly used by those opposed to experiments on animals. For example, organizations like the National Anti-Vivisection Society take their name from the word. It is ironic that, although it was the animal experimenters themselves who coined the term in 1800s, today they vociferously object to any use of the word.

What is the definition of the phrase* animal-based studies*?

The term *animal-based studies* is used to describe any research using animals, be it for experimentation or research as defined above. The term seems a good way to communicate any research done using animals, without inflaming the issue, and we will frequently use it.

Aren't you splitting hairs just a bit when you distinguish between terms like research, experimentation, and model?

No. It is important to remember that if words are to have meaning, those words must be used in a consistent manner and not misused for spin purposes. Lewis Carroll wrote in *Through the Looking Glass*:

> "When I use a word," Humpty Dumpty said in a rather scornful tone, "I mean just what I choose it to mean, neither more nor less." "But the question is," said Alice, "whether you can make words mean so many different things?"

The most important distinction here is between the words *research* and *experimentation* although later we will be concerned with the word *predict* and what exactly it does and does not mean.

In medical discussions, the term *research* refers to the attempt to find something new or confirm something else. Pretty straightforward, right? But when living subjects are involved, then the term medical *research* is reserved for attempting to find something new on the subject involved that will either be of benefit to the subject directly or at least not harm the subject.

Medical *experimentation*, on the other hand, means attempting to find something new regardless of whether that new fact will be meaningful for the subject—and usually regardless of whether the subject has given informed consent.

Here is an example of this distinction. A veterinarian conducting research on a dog that suffers from pulmonic valve prolapse (a condition in the heart) in an attempt to find something new that will help that dog, is conducting medical research. The same veterinarian inducing pulmonic valve prolapse in a dog not otherwise suffering from the condition in an attempt to find something new is conducting experimentation. Similarly, using animals as models for humans is an example of experimentation, not research, since the animals being used do not stand to gain any benefit from it, and are likely to be harmed in the process.

The distinction between the words *research* and *experimentation* are the same, whether we are talking about animals or humans. An example of human experimentation is when a new drug is first tested on humans. It is usually tested on paid volunteers who have given informed consent. These volunteers do not need this new drug and might even be harmed by it. Nevertheless, they choose to allow themselves to be experimental subjects—usually for money. This is a case where human experimentation is considered ethical. Most of the time, though, human experimentation—such as that which took place in Nazi Germany and Tuskegee, Alabama—is unethical.

Researchers who use animals as experimental subjects often interchange the terms *animal research* and *animal experimentation* in a linguistic slight of hand also known as the fallacy of equivocation. But those same researchers quickly correct anyone who confuses *human research* with *human experimentation.*

When did the practice of using animals to learn about humans begin?

The documented use of animals to gain insights about humans began around 2,500 years ago. The process continued and gained momentum in the 1st century CE when a Greek physician named Galen used animals to study anatomy and extrapolated the results to humans. Galen combined anatomic and physiologic data from animals with his personal observation of humans to forge broad theories (actually hypotheses) of physiology. Science being in its infancy at the time, many of these fell short of accuracy.

Now known as the father of vivisection, Galen may well have never used animals in the first place, had he followed in the footsteps of Hippocrates. It was Hippocrates who fathered the concept of clinical research in the 4th century BCE. Hippocrates taught that, by observing enough cases, physicians could predict the course of a disease, both in terms of its likely effect and vulnerable population. However, the practice of post-mortem dissections (autopsies) was subsequently stanched by strong opposition from the Roman Catholic Church. This prohibition on autopsies, along with the acceptance of Galen's mistaken conclusions, virtually paralyzed medical progress for more than a thousand years.

When did things begin to change?

In 1543, Belgian anatomist and physician Andreas Vesalius began to dissect the human body, and his observations led to the refutation of most of Galen's theories. His book, *De Corporis Humani Fabrica* or *Structure of the Human Body*, was published in the same year that the Polish astronomer Nicolaus Copernicus published his own startling discoveries, including the theory that the earth revolved around the sun. Together, these two books ushered in the scientific revolutions of the 16th and 17th centuries, which set the stage for the gradual emergence of modern science. One of the ultimate fruits of these revolutions was the realization that the universe could be described in terms of natural laws that made no essential reference to God, gods, demons, or other invisible beings.

By the 1800s, the autopsy was thought to be indispensable to the advancement and practice of medicine, and it occupied center stage in the quest for medical knowledge.

What brought scientists back to using animals?

The use of animals in science was brought into the current age during the 1800s, most prominently by Claude Bernard, the French physiologist who founded modern experimental physiology. Bernard was one of several investigators of the time who wanted to make a science out of medicine on a par with Newtonian physics. A noble goal.

Because so little was known at the time, Bernard went to work using animals to learn very basic things about living beings. On the gross level—for example, the level of organs such as the liver, kidneys, and heart, the level visible to the unaided eye—animals and humans do have many things in common. The dissection (and vivisection) of animals was an effective way to learn about some of these matters. Animals could be dissected, alive or dead, and many of the things learned could be applied to humans.

Bernard's own contribution to medical science lay not in using animals to predict drug and disease response but rather in using animals to implement a scientific approach to medical research—that of experimental medicine. In this context, the gold standard is that of the tightly controlled experimental study. Bernard's work laid the foundation for the conduct of modern biomedical research. And it works. The use of tightly controlled experimental studies on mice can reveal much about the biology of mice.

So what was the problem with Bernard's methodology?

There was no problem with Bernard's methodology—just his assumptions, because he believed that such experimental animal studies had direct application to human biology. In particular, he believed that nonhuman animal studies could be used to predict human biomedical outcomes. These matters are far more complex than Bernard could have dreamed in his lifetime. From a scientific perspective, Bernard has been treated too harshly by many in the animal rights movement. His science was sound at the time; however, his ethics were questionable even by the standards of his day. Upon finding Claude vivisecting the family dog, his wife and daughters founded one of the early anti-vivisection societies. One purpose of this book is to contrast science and ethics. Bernard made some mistakes scientifically speaking but he made some conceptual breakthroughs as well. Whether what he did was ethical is a debate for another book.

During this time, when Bernard was popularizing the use of animals in science, did anyone have any objections to the practice?

To answer this question properly, we need to first look at two critical periods in history that greatly influenced science and philosophy. The 1700s—the century of Benjamin Franklin, George Washington and Thomas Jefferson—was known as the Enlightenment, or the Age of Reason. During this period, philosophers believed strongly in the use of reason as the best method of learning truth. However, in the 1800s—even as Bernard was developing a scientific approach to medical research—the Enlightenment had given way to the Age of Romanticism.

During the Age of Romanticism, emotion, imagination, creativity, and intuition were valued over reason. This repudiation of Enlightenment values led to a growing movement against science in general. Many people began to look at science as the enemy—and these included some anti-vivisectionists, as well as members of the temperance and anti-vaccine movements.

The anti-vivisectionists objected to animal experimentation based on ethical grounds, but they also took it well beyond that. They sought to prove that it was not important in discovering facts about life in general and facts about humans specifically. The inherent bloodiness of animal experimentation made it a natural stepping-stone to rejecting science as a whole.

Why is this relevant to our discussion today? Surely today's anti-vivisectionists recognize the importance of science, even though their 19th century predecessors often didn't.

In fact, there are still vestiges of the anti-science movement in Europe and in the U.S. There is still a strong anti-vaccine sentiment among some people, fueled recently by the (now-discredited) belief that the preservative used in some vaccines caused autism. And there is still a great suspicion of science in general. Some refer to the so-called facts coming from such movements as junk science while others refer to the thinking of these individuals as junk thought [see Jacoby for a review of this topic [1]].

Some in the animal protection movement share this sentiment; they oppose science in general and modern medicine in particular while touting alternative medicine, chiropractic, homeopathy, and "other ways of knowing." As a result, their arguments against animal experimentation are filled with fallacious reasoning and outright errors about science and medicine.

The late Hans Ruesch was a leader of this group and made the following statements in his book *Slaughter of the Innocents*:

> Today we know that the germ does not necessarily cause the disease, and the disease can insurge without the presence of that particular germ. Bechamp was among the forerunners who attached more importance to the "soil" (the body) than to the "seed" (the germ) . . . Why does a microbe cause a disease in one organism but not in another? . . . The microbes associated with a malady may abound in the environment and be present in the human body without giving rise to symptoms [2]."

Here, Ruesch is denying the Germ Theory of Disease, which many call the greatest single contribution to medicine in human history thus far.

Ruesch also wrote in his book that gonorrhea could be treated successfully with bed rest and milk. He supported vitalism—the notion that a naturally occurring chemical cannot be precisely replicated or synthesized in a laboratory because although the chemical structure may be the same, the manufactured chemical is missing "vital" unseen immeasurable forces that occur in nature [2]. This statement from The Hans Ruesch Centre sums up the anti-vaccine position:

> However, despite the almost universal acceptance of vaccination by the public and doctors alike, in reality there is not one shred of proof to back up the claim that vaccination is either safe or effective [3].

These prognostications could be easily ignored if they were isolated instances. However, there is a strong history surrounding these positions, and even today large animal rights organizations support and promote superstition and other forms of pseudoscience and junk science in their zeal to end all use of animals in science. For example, People for the Ethical Treatment of Animals (PETA) references Ruesch's book *Slaughter of the Innocent* on their Caring Consumer website [4].

Did the early anti-vivisectionists have any impact on limiting, if not ending animal experimentation?

No. As is frequently the case when people ignore or dismiss science, they tend to be on the losing end of things. Since the 1900s, the animal experimentation industry has grown exponentially. The exact amount of money, taxpayer- and charity-derived, spent on animals in research as well as the money spent by industry is unknown. Conservative estimates, however put it into the tens of billions annually.

Much of the growth of the animal experimentation industry in the mid-1900s can be traced back to an incident in 1937 that led to the passage of the federal Food, Drug and Cosmetics Act, which requires all drugs in the U.S. to be tested on animals to prove their "safety" before they are marketed to the public.

What happened in 1937 that led to the federal Food, Drug and Cosmetics Act?

In 1937, the S.E. Massengill Company decided to create a liquid form of the drug Sulfanilamide. This drug was being used to treat streptococcal infections (strep throat). The company's chief chemist developed a formulation that used diethylene glycol as a medium for the liquid Sulfanilamide. Unfortunately, at the time no one knew that diethylene glycol is a deadly poison. No testing of pharmaceutical drugs was required at the time so the liquid Sulfanilamide was released to the public. One hundred and seven people—mostly children—died.

There was great public outrage, and to stem the sweeping tide of fear and criticism, the U.S. Congress passed the federal Food, Drug and Cosmetic Act, which is administered by the U.S. Food and Drug Administration (FDA). This law requires that all pharmaceutical drugs (as well as any chemical compound that alters the chemistry of the body) be tested on animals to prove their safety before they can be marketed to the public. We now know that safety testing on animals is not predictive for humans but animal testing seemed a solution at the time [5].

Then why do the FDA and EPA continue to mandate that new chemicals and drugs be tested on animals before a manufacturer is allowed to distribute the chemical?

Tradition. But this is starting to change as some scientists and nonscientists have started asking, along with us: "Does it work?" An April 12, 2008 front page *Washington Post* article included this passage:

> "The reason we use animal tests is because we have a comfort level with the process . . . not because it is the correct process, not because it gives us any real new information we need to make decisions," said Melvin E. Andersen, director of the division of computational systems biology at the Hamner Institutes for Health Sciences near Raleigh, N.C. "Animal tests are no longer the gold standard," he said. "It is a marvelously new world." "Some animal tests haven't changed in 60 years," said Thomas Hartung, head of the European group. "The tests are frozen in time. This is not science. Science is always moving ahead."

Scientists Wall and Shani stated: "We conclude that it is probably safer to use animal models to develop speculations, rather than using them to extrapolate [6]." In other words to seek information—not to predict human response. This is in keeping with a Memorandum of Understanding released by the national Institutes of health (NIH), and Environmental Protection Agency (EPA) in February 2008:

> Two NIH institutes have formed a collaboration with the EPA to use the NIH Chemical Genomics Center's (NCGC) high-speed, automated screening robots to test suspected toxic compounds using cells and isolated molecular targets instead of laboratory animals. This new, trans-agency collaboration is anticipated *to generate data more relevant to humans;* expand the number of chemicals that are tested; and reduce the time, money and number of animals involved in testing. Full implementation of the hoped-for paradigm shift in toxicity testing will require validation of the new approaches, a substantial effort that could consume many years. . . .
>
> The MOU and the plans articulated in the *Science* article provide a framework to implement the long-range vision outlined in the 2007 National Research Council (NRC) report, Toxicity Testing in the 21st Century: A Vision and a Strategy, which calls for a collaborative effort across the toxicology community to rely less on animal studies and more on *in vitro* tests using human cells and cellular components to identify chemicals with toxic effects. Importantly, the strategy calls for improvements in dose-response research, which will help predict toxicity at exposures that humans may encounter. [7] (Emphasis added.)

A pro animal experimentation report titled *The use of non-human primates in research 2006* stated: "It's undoubtedly the case that all animal models are limited in their predictability for humans [8]."

The National Institute of Allergy and Infectious Diseases (a division of NIH) acknowledged, at a summit they held in 2008 following the failure of a Merck AIDS vaccine in 2007, that the rhesus macaque system now used to test potential vaccines is not predictive and in fact has *not* been working out well for researchers. The Merck vaccine failed to protect against HIV infection in humans despite doing so in monkeys.

The FDA and EPA are giant government bureaucracies. As such, they are slow to keep up with current technologies. Eventually the FDA and EPA will bring their guidelines into the current millennium.

Didn't Nazi Germany ban animal experiments even while animal experimentation was growing in the United States?

No. Hermann Goering banned animal experiments in all Prussian Territories in August 1933. In 1934 the federal government of Germany banned all experiments on animals that resulted in "unnecessary tormenting or rude treatment of animals." However, Nazi Germany, like some who misrepresent science today, had a habit of using words to mean things in violation of the dictionary definition. For example, Hitler was frequently referred to as a vegetarian who liked ham sandwiches and Hermann Goering said: "Two plus two makes five if the Führer wills it." [9, 10] (For more on Nazi Germany's inconsistencies see *Hitler: Neither Vegetarian nor Animal Lover* by Ryan Berry. Pythagorean Books. 2004.)

Despite evidence to the contrary, the *Nazis outlawed vivisection and we don't want to be like them* argument surfaces from time to time and from sources that should know better[9, 11, 12].

Experiments on animals could and were done during and after the early 1930s when Nazis came to power, even after the 1934 ban [10]. Ulrich Fritzsche, M.D. wrote in 1990 in *Hospital Practice*:

> The logic of the claim perpetuated by Harold J. Morowitz that German antivivisection laws led to "cruel—often lethal—experiments on humans" ("Humans, Animals, and Physicians' Waiting Rooms," *Hospital Practice*, November 15, 1989) is difficult to maintain in view of historical facts. For example: Even a cursory review of *Pflüger's Archiv für die Gesamte Physiologie*, a major basic science journal, will demonstrate that between 1933 and 1944 animal experimentation for biomedical purposes played the same role in the German Reich as anywhere else in the world, as judged by the numbers of studies performed and species and numbers of animals involved! [10]

Fritzsche wrote again in *Anthrozoös* in 1992:

> Arnold Arluke and Boria Sax quote an AMA [American Medical Association] source "that a ban on vivisection was enacted in Bavaria as well as Prussia, although the Nazis partially retreated from a full ban" (Understanding Nazi Animal Protection and the Holocaust, *Anthrozoös* 5:6-31, 1992). Several writers have recently commented on this subject in a similar misleading fashion. Frederick Goodwin declared on the MacNeil/Lehrer Newshour that "the only people in modern society that have not used animals in research were the Nazis" (April 24, 1989). Ex-Stanford President Donald Kennedy has been particularly vocal, going so far as to remind people that antivivisection was one of the policies of the Hitler regime (Holden 1989). Arthur Caplan, Larry Horton, and Harold Morowitz repeat variations of these inaccuracies (Caplan 1990, Horton 1988, Morowitz, 1989). It appears that, in the heated debate over the issue of animal rights, the need for critical analysis of the true historical facts regarding animal experimentation in Nazi Germany has been lost.
>
> On January 6, 1934, a translation of recently issued German regulations for the conduct of animal experimentation appeared in the British journal *The Lancet* with the comment: "It will be seen from the text of these regulations that those restrictions imposed [in Germany] follow rather closely those enforced in this country [England]" (Anon. 1934) . . . One can easily extract one study using large numbers of dogs from the Archiv for every year between 1933 and 1944, despite Hitler's "fondness" for dogs. Gerhard Domagk (1935) tested the first sulfonamide antibiotic on cats, rabbits, and mice. Rhesus monkeys were used in 1940 in cardiovascular experiments (Deppe 1940).
>
> Whether or not the Nazis practiced vivisection is primarily of historic interest but has no relevance to the current debate over the ethics of animal experimentation. Too often the myth that the Nazis did not practice vivisection is used as a smokescreen to discredit those who argue against the use of animals in research today. [13] (See *Anthrozoös* for references referred to.)

But whether Nazis used animals or not is ultimately immaterial to whether the animal model is scientifically viable. It is yet another example of fallacious reasoning to juxtapose Nazi Germany next to something and conclude that the subject in question is bad because the Nazis did it. The Nazis consumed water, food, and air, wrote with pencils and pens, and drove cars. None of these activities are evil just because the Nazis did them.

Why do the Nuremberg Code and Helsinki Declaration require animal testing?

Various declarations and laws were made in response to the world's horror upon discovery of Nazi Germany's human experimentation and the Holocaust. Like the federal Food, Drug and Cosmetic Act, which was a reaction to the tragic death of the innocent people who consumed liquid Sulfanilamide, these efforts were well intentioned. However, they have not stood the test of time, as they are based on the belief that animal models were predictive for humans.

You noted earlier that the animal experimentation industry has grown exponentially since the mid-1900s. How many animals are used in science in the United States each year?

The exact number of animals is unknown. All estimates are just that—and they vary greatly. Madhusree Mukerjee, former editor of *Scientific American* writing a book review in the August 2004 issue of *Scientific American,* stated:

> In truth, animal welfare legislation and public concern are both more focused on pain than on death itself. Philosophically, the "cost" of death hinges on the worth of an animal's life. Anyone who has tried to stomp on a cockroach will have gained the impression that even such a lowly creature cherishes life. But how does one measure this value? The question has become critical with a recent explosion in the numbers of *transgenic mice--close to 100 million are consumed a year in American labs alone.* (Emphasis added.)

In 2000, the federal government needed to estimate the number of rats, mice, and birds that would be included under a proposed revision to the Animal Welfare Act (AWA). They asked the Library of Congress via the United States Department of Agriculture's Animal Plant Health Inspection Service to make such a report [14]. The report concluded that over 500 million animals would be added to the AWA if mice, rats, and birds were counted. This number excluded animals in pet stores but included the use of animals in areas other than science. Science however, appears to account for the vast majority.

The report also did not count out animals used in science but already counted under the AWA, such as dogs and monkeys. In our opinion, the number of animals left out—the dogs, monkeys, and so forth—more or less equals the rats, birds, and mice included but used outside of laboratories.

To the best of our knowledge, very few rats, birds, and mice would be used outside of scientific pursuits. Pet stores would account for some, but those animals were excluded from the report. Birds as food animals (chickens and turkeys) would lie outside this count, as would birds that are hunted. Compared to science, that leaves essentially no rats, birds, and mice used elsewhere. Fish used in labs were not counted in this estimate either. Fish alone would probably account for far greater numbers than the rats, birds, and mice used for nonscientific purposes.

Regardless, this 500 million number could be on the high side. Many animal protection groups and others continue to cite an old figure for the number of animals used. The American Medical Association places a limit at 16 to 20 million animals used in the U.S. per year. The Humane Society of the United States says 25 million vertebrates are used annually [15]. This is probably a low estimate. The Dr Hadwen Trust and the British Union for the Abolition of Vivisection estimated in 2008 that 115 million animals were used worldwide in research but admitted their numbers could be low [16, 17].

The truth probably lies somewhere between the high estimate of 500 million and the three lower estimates but exactly where, who knows? The matter is difficult to settle definitively, since the number of animals used in science is not open to direct public scrutiny and because the majority of animals (which are rats and mice) are not counted. In any event, we are evidently looking at a figure of many, many millions.

The number is also difficult to settle because of the way many researchers count the animals they use. The counting is largely done on the honor system and there are many ways to get around reporting all the animals used. Some scientists count only the animals used in the final stages of the experiment. For example, say the researchers use 100 animals in the first stage of the experiment but only 80 will give the result the researchers are looking for because of technical difficulties with the experiment. If they can, they will report only 80 animals thus used. The extra twenty disappear from the records to be used again or, if the study was too invasive, they would simply be euthanized. In some situations this is difficult to manage but in others it is routine.

Some studies are very difficult to perform, and, for every one animal for which the surgery is successful, the researcher may have needed to perform the surgery on 20 animals before achieving that one successful result. The controversy surrounding animals in science is well known in the research community and everyone wants to keep the numbers down. Not many people will discuss this and most will deny it. But having worked in labs using animals and having had colleagues in universities who were truthful about the practice, we can state with certainty that the practice does occur.

Regardless of the number of animals used, the question we are concerned with remains, "Is such use scientifically justified?"

How much money are we talking about?

Animal testing, experimentation, and use of animals as models of humans is a multi-billion dollar business. Universities, individuals, animal breeders, suppliers of cages and equipment and more, all profit. Below are the costs of some items, as described by Pennisi writing in *Science* [18]:

1. Rat Stereotaxic Instrument	$4,500.00
2. For cats and monkeys	$7,215.00
3. Metabolic gas monitor	$27,300.00
4. Flat treadmill for rodents	$9,600.00
5. Incapacitance Analgesia Meter	$7,300.00
6. Sliding microtome	$9,975.00
7. Muromachi microwave fixation system for humane sacrifice with immediate deactivation of brain enzymes	$70,200.00
8. Stereotaxic device for dogs	$8,580.00

According to the same article, mouse sales amounted to over $200 million in 1999. Charles River Laboratories of Massachusetts alone sold over $140 million of animals in 1999. Experts estimate that Harlan Sprague Dawley of Indianapolis sold over $60 million in animals in 1998 and Taconic $36 million. TJL, a not-for-profit taxpayer funded corporation, sold $29 million worth of mice alone. Mice with specific genes missing cost from $100 to $15,000 a piece.

In 2005, it was observed in *Nature Medicine* that:

> Making a new knockout line is not easy: it can take up to a year and cost up to $100,000, notes Muriel Davisson, director of genetic resources at The Jackson Laboratory in Bar Harbor, Maine, one of the three NIH mouse banks. Buying a mouse line from a company is no cheaper. For instance, California-based Deltagen charges $26,200 for two pairs of live knockout mice. Additional embryos, sperm and embryonic stem cells can run up another $15,000. Even then, the mice often come attached with intellectual property strings. Company employees sometimes co-author papers and some companies demand royalties on any discoveries or products. [19]

The percentage of NIH funding which goes to animal-based studies is largely unknown. On the next page is a table from a report from the National Research Council 1985 (table 1.1) [20]. Note that mammals alone amounted to roughly 45 percent and if that number is combined with a fraction of the category that includes nonmammalian vertebrates, such as birds and reptiles (other), the number easily goes over 50 percent.

This is an old table but is still the most recent record of the amount of money spent by NIH on experiments involving animals. NIH and other institutions have refused numerous requests to repeat or initially conduct such tabulation despite the fact that with today's computers such a table would essentially make itself. The percentage of money has probably stayed the same or, increased. Since 1985 basic research has if anything become more connected with animal-based research, which receives the lion's share of grants. (See next chapter for a more detailed explanation of basic as opposed to applied research.) But make no mistake, the NIH primarily gives grant money to researchers

Table 1.1

Table 4-1. Distribution of NIH Support of Extramural Research Among Humans, Laboratory Mammals, and Other Research Subjects, Expressed as Percentages of Total Dollars and of Total Projects and Subprojects[a]

Subject	Fiscal Year	Extramural Research Dollars, %	Total Projects and Subprojects, %
Humans	1977	27.5	32.4
	1978	26.8	31.2
	1979	26.8	29.2
	1980	25.0	28.9
	1981	23.8	29.7
	1982	23.2	31.5
	1983	22.9	32.2
Mammals	1977	43.5	41.9
	1978	44.0	42.5
	1979	44.9	43.8
	1980	45.0	44.2
	1981	47.3	44.1
	1982	48.1	43.5
	1983	47.9	42.7
Other[b]	1977	29.4	25.6
	1978	29.3	26.3
	1979	28.2	27.0
	1980	29.8	26.9
	1981	28.9	26.0
	1982	28.7	25.0
	1983	29.2	25.1

[a] Unpublished information provided by Division of Research Resources, National Institutes of Health.

[b] This category includes invertebrates, nonmammalian vertebrates, bacteria, viruses, mathematical and computer simulations, and other subjects.

performing basic research.

Nathan and Schechter in *JAMA* 2006:

> In 1995, Harold Varmus, then director of the NIH, became concerned that the NIH was not providing sufficient support for clinical research. He formed a panel of experienced academic leaders who accomplished 3 important goals.3 First, the panel defined clinical research broadly and estimated that about *one third of the NIH extramural budget was devoted to clinical research* [that would mean roughly two-thirds would go to basic research]. That amount seemed reasonable to the panel, considering that much basic research would be expected to be required in such a scientific program because most basic research provides only a small piece of information that, by itself, cannot be used in the clinic without further information. But the panel cautioned the NIH not to let the ratio decrease further.3 [21] (Emphasis added.)
>
> Reference
>
> 3. David G. Nathan, MD; for the National Institutes of Health Director's Panel on Clinical Research. Perceptions, Reality, and Proposed Solutions. *JAMA*. 1998;280:1427-1431.

The 1998 Nathan article referenced above also said:

> The proportion of investigators applying for clinical research grants from the National Institutes of Health (NIH) who are physicians has declined from 40% 30 years ago to 25% today.

Physicians are more likely to perform clinical or applied research than basic research, thus the above quote indicates a dwindling percentage of people even applying for grant money for clinical or applied research. If the pot of money dedicated to basic research is greater than the pot for clinical or applied research, that is where most people will go. The odds of getting the grant are simply better.

Connecting basic research on animals to possible cures for humans is stock and trade for many researchers. Freeman and St Johnston in 2008:

> Many scientists who work on model organisms, including both of us, have been known to contrive a connection to human disease to boost a grant or paper. It's fair: after all, the parallels are genuine, but the connection is often rather indirect. DMM [the journal *Disease Models & Mechanisms*] is about something quite different. This new journal is aimed at people who set out with an explicit goal to investigate human disease using model organisms. [22]

With roughly two-thirds of research dollars going to basic research and most basic research using animals we can safely confirm the 1985 numbers of at least 50 percent of research dollars going to animal-based studies. If we assume a $30 billion dollar NIH budget, one can estimate tens of billions annually are spent on animal-based studies from NIH alone. As we have stated, the total amount of money, taxpayer- and charity-derived, spent on animals in research is unknown but conservatively measured in the billions annually. And this does not even consider money spent on animals-based research by industry.

With such an enormous amount of money involved, are there mechanisms in place to ensure proper oversight?

Some have said the medical research and medical care systems are a microcosm of society itself. The problems society faces in general have their origins in the same flaws we see in the medical research and care systems: greed, corruption, special interest control of government, and an old buddy system that gives priority to the *who* (friends and contributors) not the *what* (competence).

But we can be a little more specific. In his last speech as president, Dwight D. Eisenhower warned that the military-industrial complex was exerting too much influence in America's politics. The phrase *military-industrial complex* has been around ever since and everyone understands what it means. What has been forgotten about Eisenhower's speech that day was that he had a similar warning about the influence the government had on scientific research. There is some degree of truth to the old adage, he who pays the piper calls the tune!

For decades, research grants from the government have been more about whom you know and less about what you know or what you are doing. One Congressman stated at a hearing in 1988 that the granting system is: ". . . an old boy's system where program managers rely on trusted friends in the academic community to review their proposals. These friends recommend their friends as reviewers . . . It is an incestuous 'buddy system [23].'"

At a similar hearing in 1989 another stated: "It appears that the [medical establishment] system has changed from one of NIH giving grants for scientific research to one of scientific research being done solely to get NIH grants [24]." Though regrettable, this is an understandable state of affairs given that academic careers, for example, often hinge on the ability to bring in grant money. Meyers wrote the following in 2007 in *Happy Accidents: Serendipity in Modern Medical Breakthroughs*:

> About 90 percent of NIH-funded research is carried out not by the NIH itself on its Bethesda campus but by other (mostly academic medical) organizations throughout the country. The NIH gets more than 43,000 applications a year. Through several stages of review, panels of experts in different fields evaluate the applications and choose which deserve to be funded. About 22 percent are approved for a period of three to five years. The typical grant recipient works as a university and does not draw a salary but is dependent on NIH funding for his or her livelihood. After the three- to five-year grant expires, the researcher has to renew the funding. The pressure is typically even greater the second time around, as the university has gotten used to "absorbing" up to 50 percent of the grant money to use for "overhead," and by now the scientist has a staff of paid postdocs and graduate students who depend on the funding, not to mention the fact that the continuation of the scientist's faculty position is at stake. [25]

George A. Keyworth II, White House science adviser in the Reagan Administration, stated in a congressional hearing: “American science has become a bureaucracy. As with all bureaucracies, preserving the status quo has become the overarching goal, replacing the pursuit of excellence [26]."

The problems with the granting process are not confined to animal-based grants. For example, organizations such as NASA also struggle for access to increasingly scarce research funding.

Given the vast resources that have been invested, has anything good come from animal experiments?

If by *good* you mean has science advanced or has society learned more about life in general, then the answer is an unconditional *yes*.

If you had no knowledge of the mechanisms of life and wanted to learn as much as possible using only animals, you could replicate the history of biology up to and in some cases including the knowledge we have today. However, in order to learn about specifically human conditions, you would have to study humans.

Nevertheless, a vast majority of life is nonhuman. At the level visible to the naked eye (what we and others call the *gross level*) or even through a light microscope, all life has much in common. For example, if you wanted to see what the building blocks of anatomy are you could look at tissue from any animal or plant and find cells. If you wanted to know how nutrients and oxygen are distributed to the tissues, you could study any mammal and find that the heart circulates blood through arteries, capillaries, veins, and so forth.

Some maintain that society could also have arrived where we are today without ever using animals. Clearly, one could have used exclusively human tissues and discovered the basic biochemistry and anatomy of life. By studying humans and only humans, one could also have made crucial discoveries in surgery and pharmacology. (Sadly, advances in surgical technique are all too often the result of lessons learned on the battlefield.)

We could engage in a long, and in our opinion pointless, debate about what discoveries were dependent upon using animals and what discoveries were not. Reasonable people will acknowledge that animals were used in the past to make important discoveries about life. Whether one agrees with animal experimentation or not, one should not fool oneself about its place in history. The animal protectionists who deny the value of science would serve themselves well by learning this important lesson. There are those in this movement who deny that HIV causes AIDS, believe astrology over astronomy, and who seek a “natural healer” instead of a physician when they’re feeling ill. Moreover, they deny that anything has ever been learned from animal experimentation (or, indeed, *any* experimentation). This is the danger of people who do not understand critical thinking and consequently reject science (see later chapters for more on this). People cannot embrace science and critical thought when it suits them only to dismiss it at other times.

It was profoundly disturbing, for example, to hear well-known comedian, political pundit, and self-described animal protectionist Bill Maher echo Hans

Ruesch's beliefs to a hearty round of applause from the audience on his show *Real Time* on March 4, 2005:

> Maher: I don't believe in vaccination either.
> Healy [Bernadine]: Oh, dear.
> Maher: That's — what? That's another theory that I think is flawed. And that we go by the Louis Pasteur theory even though Louis Pasteur renounced it on his own death bed and said, "Beauchamps was right; it's not the invading germs, it's the terrain. It's not the mosquitoes, it's the swamp that they're breeding in." [applause]

Let there be no doubt that society can learn important scientific lessons by experimenting on animals. Let there be no doubt that society can learn important scientific lessons by experimenting on Jews in concentration camps and black men in Tuskegee, Alabama. The fact that science can advance in these ways has no correlation to the ethics surrounding the endeavor. Has the good that has come from animal experiments outweighed the ethical objections to using animals? That is a philosophical, not a scientific question and we will not be addressing it in this book.

The real question we need to be asking is, what are the ways animals can and cannot be used in science? Can animals be used to predict human drug and disease response? Our answer is no, for reasons that we will explore later in this book. The fact is, the majority of animals that are used in testing and research are used in the hope that they will predict human response. This is a false hope. It does not work. At the same time, there are many scientifically valid ways that animals are used in science, which we will discuss in greater detail in the next chapter.

Chapter 2. The Scientifically Viable Uses of Animals

What are the ways in which the use of animals is scientifically valid?
There are at least seven ways (numbers 3-9 in table below) in which the use of animals is scientifically valid:

1. Animals are used as predictive models of humans for research into diseases such as cancer and AIDS.
2. Animals are used as predictive models of humans for testing drugs or other chemicals.
3. Animals are used as "spare parts", such as when a person receives an aortic valve from a pig.
4. Animals are used as bioreactors or factories, such as for the production of insulin or monoclonal antibodies, or to maintain the supply of a virus.
5. Animals and animal tissues are used to study basic physiological principles.
6. Animals are used in education to educate and train medical students and to teach basic principles of anatomy in high school biology classes.
7. Animals are used as a modality for ideas or as a heuristic device, which is a component of basic science research.
8. Animals are used in research designed to benefit other animals of the same species or breed.
9. Animals are used in research in order to gain knowledge for knowledge sake.

In this chapter, we will discuss these modalities further and explain why they are scientifically viable.

What do you mean by using animals as spare parts?
It is the process of repairing defective human tissues or organs by using tissue or organs taken from an animal. One of the most successful examples of using animal tissue in this way is in aortic valve (a valve in the heart) replacement. Replacing a defective aortic valve with either a pig valve or a cow valve (mostly pig valves are used) is the most common tissue valve-type operation in heart patients. It is called a *heterograft*. Today, many valves have both a pig component and synthetic (nonanimal) component.

Replacement valves can also come from humans who have donated their hearts. These are called *homografts*. Mechanical or synthetic valves are the third type. There are advantages and disadvantages to each, but all are viable and safe for patients.

Burn patients are also recipients of pig parts, namely skin, although replacement for the burned skin usually comes from the patient himself. An instrument is used to cut away a thin slice of skin from an unburned portion of the body, say for instance the thigh. This split thickness skin graft is then placed onto the burned area where it will grow to cover the burn. If skin from the patient is not available, then skin from a cadaver or pig can be used in order to provide temporary covering.

Aren't there grave risks associated with putting animal tissue into humans?

Yes, whenever tissue from any foreign source is introduced into a human, there is a risk for transmission of viruses and other disease-causing agents, as well as undesirable host-immune responses. But this does not mean, when all reasonable precautions have been taken, that foreign tissue should never be implanted, as the case of pig valves clearly demonstrates. One reason pig valves are safe is that the valve tissue is dead whereas the use of living tissue risks transmission of retroviruses and so forth.

The risks associated with implanting living tissue across species is one of the many concerns surrounding the issue of *xenotransplantation*—that is, animal-to-human living organ and tissue transplants. Taking living cells from pig pancreases, for example, and implanting them into people with diabetes, or taking an intact liver or heart and implanting it into a human would harbor the risks of viral and other disease-causing transmissions, not to mention undesirable immune responses.

How are animals used as "factories" for human benefit?

One might describe the harvesting of insulin as one of the most well known examples of using animals as factories. Insulin is a hormone that is used to treat people suffering from diabetes. Until the early 1980s, insulin for diabetics was obtained primarily from the pancreases of cows and pigs as a by-product of the slaughtering process. For decades, insulin harvested from slaughtered cows and pigs saved human lives. Today, however, insulin is manufactured by genetically engineered bacteria.

Another example of using animals as factories can be found in the case of respiratory distress syndrome (RDS). RDS is a disease suffered by very premature infants and is caused by a lack of surfactant in the lungs. One way of treating RDS has been to administer surfactant obtained from animals.

What is a bioreactor, and how are animals used as bioreactors for human benefit?

Simply defined, a bioreactor is a vessel (artificial or natural) or apparatus in which microorganisms can be maintained in a liquid environment and used to perform chemical reactions. Think of a bioreactor as a kind of incubator for growing cells, viruses, and other microorganisms.

An example of using animal cells as bioreactors is the therapy known as monoclonal antibodies. Monoclonal antibodies (mAbs) are antibodies that come from one type of immune cell and are used to treat diseases like cancer. The original research on mAbs was performed on mice, and mice were initially used to produce the mAbs given to humans.

How are animals and animal tissue used to study basic physiological principles?

There is still much we do not know or understand about the basic animal body. By studying animals, especially less complex animals like fruit flies and worms, we can and have made advances that relate to humans and human health.

One of the best examples of this is the emerging field of evolutionary and developmental biology (Evo Devo), which has relied heavily on the fruit fly. Sean Carroll wrote in *Endless Forms Most Beautiful*:

> Almost immediately after the first sets of fruit fly genes were characterized came a bombshell that triggered a new revolution in evolutionary biology. For more than a century, biologists had assumed that different types of animals were genetically constructed in completely different ways. The greater the disparity in animal form, the less (if anything) the development of two animals would have in common at the level of their genes . . . But contrary to the expectations of any biologist, most of the genes first identified as governing major aspects of fruit fly body organization were found to have exact counterparts that did the same thing in most animals, including ourselves. This discovery was followed by the revelation that the development of various body parts such as eyes, limbs, and hearts, vastly different in structure among animals and long thought to have evolved in entirely different ways, was also governed by the same genes in different animals. The comparison of developmental genes between species became a new discipline at the interface of embryology and evolutionary biology—evolutionary developmental biology, or "Evo Devo" for short. [27]

It is important to point out here that scientists used relatively simple organisms like fruit flies and worms to study these basic principles. It would have been difficult, if not impossible to study these principles in more complex organisms like mammals. What we are calling *relatively simple organisms* should perhaps be called *less complex* because even the fruit fly is a very elegant piece of work, which human bioengineers are far from reproducing. But compared to mammals, fruit flies and worms are simpler. By using such organisms, scientists can study very basic evolutionary processes. This cannot be accomplished using species that more recently separated and that are more complex.

What is another example of studying animal tissue to learn about basic physiological processes?

In 1998, Robert Furchgott, Louis Ignarro, and Ferid Murad shared the Nobel Prize in Physiology or Medicine for discovering that the body uses nitric oxide gas to make blood vessels relax and expand—a discovery they made by studying tissue from pigs.

Nitric oxide (as distinguished from nitrous oxide or laughing gas) is a signaling substance found in our bodies. It directs blood vessels to dilate and in turn lowers blood pressure. The finding has already led to high blood pressure treatments, treatments for premature babies, and drugs like Viagra.

Can animals be used successfully in education?

Yes. Animals can be used successfully to demonstrate basic anatomical and physiological principles for students. For decades, students in pre-university education have dissected starfish, frogs, cats, fetal pigs, and so forth in order to formulate an understanding of basic anatomical principles. These principles include the fact that veins and arteries and nerves usually run in close proximity, that structure determines function, and that, while different tissues are constructed for different purposes, they all use the same building blocks.

How are animals used in education at the university level?

The tradition of dissection has continued into college with physiology students experimenting on rabbits and mice in order to see changes in heart rate, respiratory rate, urinary output, and so on. Medical students have used dogs, pigs, and other animals to study physiology and pharmacology, as have veterinary students. Departments of surgery have also used animals, such as dogs and pigs, to aid residents in learning how to suture.

How are animals used in surgical training?

If a surgical resident wants to learn a technical skill, such as how to sew a vein to a vein or an artery to an artery (called an anastomosis), she can practice on a renal vein from a dog or in a dog, and that skill can be learned at least in part in this fashion.

Some surgeons conduct trials on pigs and other lab animals before performing the new surgery on humans. But the practice is not without its pitfalls. The field of neurosurgery offers an example. Extracranial-intracranial (EC-IC) bypass procedures for inoperable carotid artery disease were tested and perfected on dogs and rabbits. Neurosurgeons performed thousands of EC-ICs before it was discovered the operation did more harm than good. More patients died or suffered strokes because of the operation than were saved as a result of it. This is the problem when using animals to predict human response. This was not merely using animals and animal parts to learn a technical skill.

How can animals be used as a modality for ideas?

Sometimes when a researcher is studying an animal or tissue derived from an animal, he gets an idea about something in humans. In this way, the animal is being used as a heuristic device. Sometimes the idea is similar to what the researcher is studying in animals, and sometimes it is rather far removed. Regardless, the atmosphere of doing biomedical research in animals or in some other way, lends itself to such thoughts. Human cognitive creativity can be induced in a variety of ways.

Some scientists would take this concept further. LaFollette and Shanks have discussed the use of animals as heuristic or hypothetical animal models (HAMs):

> Scientifically legitimate HAMs are not merely psychological causes which serendipitously prompt a scientist to make a discovery. As a matter of fact, a scientist might gain important insights about the metabolism of phenol after jogging a mile, listening to Beethoven's Fifth, or drinking a cup of coffee. But that does not mean jogging, listening to music or drinking coffee is the same as studying a HAM. There is no particular reason why these activities prompted the scientific insight; nor do we have any reason to think they would prompt important insights by other scientists. These are merely unique psychological causes, not scientific devices. On the other hand, scientists plausibly assume that experiments on animals can suggest fertile hypotheses about biomedical phenomena. A HAM is valuable in as much as there are demonstrable functional similarities between the model and item modelled. Since there are demonstrable functional similarities between humans and our close biological relatives, biomedical scientists infer that the results of tests in animals will probably prompt ideas about how to think about and understand the functionally analogous human phenomenon [28].

Another variation on this theme is typified by the torpedo fish and Alzheimer's disease (AD). The protein alpha-synuclein, which is involved in AD, was discovered in the torpedo fish at Stanford University in 1988. (The gene that codes for it was isolated from humans. The mutated forms of alpha-synuclein and the protein tau were originally discovered in humans with Alzheimer's and Parkinson's disease (PD). The causal link between the abnormal proteins and the diseases of PD and AD were also of human origin.) When scientists started looking for a protein that had the characteristics they found in people suffering from AD, they searched a database and found the fish data. Obviously the protein could have been characterized without fish; it could have been taken from humans and chemical analysis performed as it had been performed with the protein from the fish. But this example does illustrate how animal data can be successfully used.

This leads to our next question.

What is basic science research?

Basic science research is also called *basic research* or *pure research*. Basic research can be variously defined but it is distinguished by being curiosity-driven rather than goal-driven, which defines applied research. Attempting to find a drug to cure malaria would be applied research, not basic research. Basic researchers want to increase the amount of knowledge in the world and find value in this endeavor regardless of whether the knowledge ever leads to anything else. That being said, basic research does sometimes lead to important new drugs or other inventions. Basic biological science has traditionally studied life at the most basic level, such as what a cell is and what it is made of, what distinguishes life from nonlife, what everything is built of, and so forth.

Basic research has been very important to scientific advancement. Discoveries and inventions derived from basic science research include:

DNA
Basic biochemistry such as the Krebs cycle
The fundamental elements
The mass spectrometer
Transistors
Computer circuits
X-rays
Electrons
Nuclear power
Electromagnetic waves
Induction coils in automobiles
Global Positioning Satellite system

How does basic research differ from applied research?

Basic research stands in contrast to applied research, which is goal-oriented. In applied research, the scientists often want to make something commercially viable. There is no doubt that research is a continuum ranging from basic to applied, and it is not always easy to categorize a specific research project. But one thing remains certain. Basic research makes no claims of applicability. Arthur Kornberg stated in a 1995 editorial in *Science*: "We are urged: Do strategic basic research! Do targeted basic research! How can we make clear the oxymoronic nature of these terms[29]?" J. J. Thomson, the discoverer of the electron, defined basic research in 1916: "By research in pure science I mean research made without any idea of application to industrial matters but solely with the view of extending our knowledge of the Laws of Nature [30]."

Historically, animal use in research was synonymous with basic research. It was easy to dissect or vivisect animals without any particular end in mind. If you were curious about a phenomenon or wanted to lean more about life in general, animals could be used.

How are animals used in basic research today?

Science is still using animals successfully in basic research, but the kind of animals being used has changed.

Mice are still the most commonly used, but animals like fruit flies are allowing a greater understanding of fundamental genetic processes. (Actually, flies and worms might outnumber mice but since none are counted it is impossible to say for certain.) Other nonanimal living species like fungi are also being used to discover very fundamental life processes.

It is important for our discussion to recognize that basic science research was once the rule, not the exception. As society began to realize what tremendous breakthroughs medical science was capable of, they began to demand that government-funded researchers focus on curing specific diseases—AIDS, breast cancer, and so forth. Funding agencies were happy to oblige by creating a climate in which researchers competing for scarce resources had to go through the motions of dressing up basic research as research with practical application (something achieved if by no other means, then by strong hints of practical relevance).

This created a conundrum for the scientific community, because what many academic scientists claim they wanted to do was research into interesting puzzles, the application of such research not withstanding. What the public, and therefore Congress, wanted was cures. Consequently, scientists now tend to phrase their research grant proposals in terms of what their research might possibly accomplish, rather than why they are really doing it.

The following are some examples from grant proposals:

> Grant Number: 1R21N2052355-01A2
> An Experimental Model of White Matter Infarct: Abstract: An animal model of white matter infarct will be valuable in the understanding of the underlying mechanisms of subcortical ischemic stroke, neuronal reorganization leading to motor recovery and most importantly, the evaluation of therapeutic interventions [for humans]. The goal of this project is to develop such an animal model . . . .

And:

> Grant Number 1R21AG026482-01A1
> Experimental approaches to traumatic brain injury in aging: Abstract: . . . hypothesis that increased inflammatory responses to a comparable injury in old mice contribute to the worsened outcome . . . The overall goal of these studies is to provide a rationale for specific treatments for TBI in the elderly [humans].

And:

> Grant Number: 7R01NS)40587-05
> Lentiviral delivery of GDNF and BCL-2 in PD model: Abstract: In this regard, two experimental strategies have been demonstrated to be successful in animal models of PD [Parkinson's Disease] . . . These studies will serve as the preclinical foundation to determine whether this *in vivo* method to deliver these potent trophic factor and antiapoptotic genes will be suitable for testing in [human] patients with PD.

In such research proposals, the researchers are saying that they believe their animal models are predictive for humans. This is classically what applied research is all about. Some questions arise as a result of such proposals: First, do the scientists actually know they are doing basic science research while telling the government that the research is applied? If so, they are committing fraud (admittedly in a funding environment that is complicit). Fraud is defined by the Encarta dictionary as: 1) the crime of obtaining money or some other benefit by deliberate deception; 2) somebody who deliberately deceives somebody else, usually for financial gain [31]. Taking taxpayer money under certain conditions that one knows are not true is fraud.

Second, if the scientists actually believe their research is going to be applicable to humans, what is their basis for this? Using animal models to predict human response is not, as we shall see, a valid scientific principle. We see no way to resolve the fact that scientists are asking for money to cure diseases but claiming among themselves that such research is actually basic science research, with no goal in mind.

James W. Hicks, PhD, is Professor of Comparative and Evolutionary Physiology at the University of California-Irvine (UCI). In a debate (transcript is available at www.curedisease.com) at UCI on March 2, 2006 with one of us (RG), Dr Hicks stated:

> So now, I agree. Studying a rat will not tell you anything of specific information about a disease in a baby, specific information about the disease in a baby. So if you want to study babies, you study babies. Obviously. That's what you have to do. Will this [a rat] tell you anything about that [a baby]? Well, not really. Not really . . . So now, I agree. Studying a rat will not tell you anything of specific information about a disease in a baby, specific information about the disease in a baby. You study babies if you want to know about babies. But causal analogical models, Dr. Greek's completely right, it dominated physiology and biochemistry for 20 --for the --most of the twentieth century, most of the nineteenth century. It was the result of linear thinking, that if I study a dog, it's directly applicable to the human, if I study a rat it's directly applicable to a human, if I study a frog it's directly applicable to a human. And who was that the result of? It was a great man, Claude Barnard, French physiologist who was the sort of father of physiology of the twentieth century. He was -- did most of his work in the late 1800s. And his thinking and his students, the people that came out of his lab that were associated with, they were not evolutionary biologists. They weren't trained. Evolutionary biology was a new science itself. They hadn't even read Darwin probably. But so they produced students who thought the same way, that there was causal linkage between studying an animal and directly relating it to human disease. And this did dominate the field for a long time . . . But what happens in terms of CAMs [CAMs are predictive models] is that what -- what he's attacking is the way scientists talk to the public. And this is a problem. And so when you ask me does every experiment result in a benefit to humans, you know, every time I'm interviewed by reporters --and I've done -some of the research that I have done in my lab has had interest by the popular press. Every time they'll ask me that question, well, what benefit is it for humans? That's the common question. Well, do you ask that of a physicist who's studying black holes?

No, but physicists do not sell their research to the public promising cures for disease. We expect people who say they are doing biomedical research to find cures to actually find cures, at least some of the time. We do not expect physicists to find cures. We expect them to do something that furthers our understanding of black holes. Dr Hicks:

> Biologists are studying, trying to understand nature, how nature works, how life works. Some of the information will have benefit towards humans. A lot of the information won't have direct benefits to humans but will increase our understanding of the natural world. That's what motivates a lot of people.

And all that would be fine if in the grant proposal they said the above. Dr Hicks:

> What motivates a lot of biomedical scientists is using animal models to generate novel and new ideas. However, what has happened—and Dr. Greek is entirely correct—is that to sell the idea, quote, unquote, sell it, I would—I would agree that many biomedical scientists can be accused of over promising and under delivering. But the over promising is what's required or the—or is what they have to define why are you doing this research and how's it going to directly benefit humans?

That's our point. Dr Hicks continues:

> What scientists have not done well in the 20th century and what they don't I think do well now is explain what is the scientific process? Where do ideas come from? Do they just come out of the air? Are you sitting here one day and suddenly go, "Oh, that's a great idea. I will I'll use that. Yeah, that's—that's how the cell cycle works. That's fantastic. Now I know where it's come from"? No. They do experiments.

We agree, but the issue here is experimentation on animals. We completely agree that experimentation *per se* leads to new knowledge. The question is, if we all agree that animals are not predictive, is using animals, for example mammals, in basic science research a good use of resources or are there better, less controversial ways? Dr Hicks:

> And many experiments lead nowhere. But in defending [what] they're doing--the experiment—they often will tell the public or tell the granting agency or tell the Congressman the end point, "I'm doing—I'm studying cell cycling because it's going to cure cancer." Well, you know, cell cycling, how cells determine how they're going to divide, that is important to cancer biology. But will that specific project cure cancer? Well, probably not. But it will—might lead to some new insights into the cell cycle, which then later on might lead into some additional insights into the cell cycle, which then might lead on—and this can take a long time.

Again, Hicks makes our point for us. We have accused researchers of being disingenuous, and he concurs.

> So I do think scientists are guilty of over promising and under delivering. And they shouldn't be excused from it. What I tell my students, what I tell my postdocs is that when you're asked that question do not fall in the easy trap, "Oh, I'm studying this because it's really important for heart biology." Tell me why you're studying it. What excites you about the question? What is—why are you really doing it? And if they want to continue to write something about how it's going to help heart biology, that's the reporter's problem, not the scientist's problem . . . And yet he said, you know, scientists will say that and then they wink, wink, don't—then they explain this CAM. Well, no, it's just they have fallen into the trap themselves of always trying to give the ends point and not talking about the pathway to—to ultimate understanding. Because people don't

> want to hear about the pathway. It's too long. It takes 50 years to get there, not two years . . . .

And we must judge those 50 years in light of the other options we have available; other areas of research we could be funding instead, such as areas like human-based research and research using less complex organisms. Basic science research using rodents and dogs and monkeys must be judged against all the other research options out there. The vested interest groups make it sound like it is animal-based research or nothing. This is an example of the fallacy of the false dichotomy. Dr Hicks:

> So I think it depends on the question. I do agree with Dr. Greek that—that if someone directly thinks that there is a—that they're using a CAM—a real model, that they're studying a rat, that's really going to cure hypertension, I think that that's bad science. But I just don't—the scientists that I interact with, the ones that I discuss things with, the ones that I talk with, they don't think that way.

They probably don't. That is why they are disingenuous. They think one thing and say another.

Can animals be used to obtain knowledge for knowledge sake?

Of course. Any time good scientific research is performed, more data are added to the world of science. There are those in the animal experimentation community who are very up front about doing research not for any goal—whether long- or short-term—but simply to add knowledge to the world, regardless of whether that knowledge will ever be used for anything.

What about studying animals in order to benefit other animals?

Animals can—and should be—used in research for animal diseases and drugs used in veterinary medicine. There is no substitute for what is known as clinical research. However, it must be emphasized that research in dogs will be more reliably applied to dogs than cats. In other words, it must be species-specific, so as not to repeat the huge mistake of applying animal-based studies to human beings. It should actually be breed-specific and if possible tailored to the individual animal. Obviously, if one wishes to cure cancer in Boxer dogs, studying Boxer dogs as opposed to Great Danes will aid in this endeavor. No two individuals—be they two dogs, two cats, or two humans—have exactly the same genes, gene expression, and regulation. This is true even of monozygotic (identical) twins. So when possible, medicine should be gene-based, not breed- or species-based.

Some suggest that the way human research is conducted can inform animal research. Before commencing any research on the species in question, one would perform all relevant pre-clinical studies. This would include *in vitro* research, which involves cell and tissue culture. (Ideally, one would study the cells of the animal species in question.) Other pre-clinical studies would include computer modeling, and all other minimally invasive methods of research, which would yield as much information as possible. The scientist could proceed to the individual in question just as is done in human research.

Just because there are scientifically viable uses for animals in science, as you say there are, does that mean using animals is always the best way?

Not necessarily. As we discussed earlier, the original research on mAbs was performed on mice, and mice were initially used to produce the mAbs given to humans. However, this approach was not without its problems. The mouse mAbs did not work well in humans because they did not trigger the appropriate response. Additionally, the mouse mAbs had a very short half-life in humans.

Today mAbs that are almost 100 percent human-derived are available that circumvent all these problems. Loisel et al. in 2007:

> Animal models are not suitable for predicting the immunogenicity of therapeutic mAbs in humans, and transposition of the immunogenic potential of therapeutic antibodies in animals to the human situation has no scientific rationale, even in primates . . . In conclusion, complement activation by therapeutic antibodies in animal models is strongly influenced by a variety of parameters and does not necessarily reflect the human situation [32]. . . .

What are the risks in using animal-derived ingredients in in vitro tests?

First, understand that a great deal of *in vitro* research involves animal-derived products like blood from cows and cells from other animals in the growth medium. Consider fetal bovine serum (FBS), which is used in many cell cultures. There are problems with using FBS:

1. There is batch-to-batch variability, which is not a good thing in a culture medium that by definition needs to be consistent.
2. The amount and type of growth factors and growth inhibition factors varies from batch to batch.
3. The FBS may vary, depending on the sex of the fetus, species, and developmental stage.
4. FBS "can interfere with genotypic and phenotypic cell stability and can influence experimental outcomes."
5. FBS can "suppress cell spreading, cell attachment, and embryonic tissue differentiation."
6. FBS "can be contaminated with viruses, bacteria, mycoplasma, yeasts, fungi, immunoglobulins, endotoxins, and possibly, prions."
7. Many substances in FBS have not been identified, and the effect of these substances on cultured cells is unclear[33].

Historically, we have seen the risks of contamination in the development of vaccines: Monkey kidney cells that were used in the production of the original Salk and Sabin polio vaccines were contaminated with Simian virus 40 (SV40), a virus that infects several species of monkeys.

Another serious problem with using tissue from animals is the risk of infectious disease.

How can these problems be resolved?

In all likelihood, most of the animal components currently used in *in vitro* research could be replaced with human- or synthetic-derived products. For example, synthetic versions of the animal surfactant used to treat very premature infants with respiratory distress syndrome have now become available.

Consider, as another example, the insulin assay, which is used to ensure that the insulin is pure and of the appropriate concentration. Historically, this assay used antibodies to detect the insulin, and these antibodies were being produced from cells that had been placed into the abdomens of living mice. The Physicians Committee for Responsible Medicine (PCRM) recently introduced an insulin assay that does not involve animals or animal parts. PCRM:

> They [laboratories] used antibodies to detect insulin, and these antibodies were produced from cells that had been placed into the abdomens of living mice. The unfortunate animals become painfully swollen with antibody-filled fluid, which the laboratories extract with a needle and use in test kits. Considered "living factories," these animals are used by the millions each year, not just for insulin assays but for all types of medical tests. Rather than support this form of animal use, PCRM decided to look for a lab that could grow the antibody-producing cells in a test tube. Working under Dr. Barnard's direction, PCRM research analyst Megha Even, M.S., took on the challenge. She soon located a laboratory in Emeryville, California— BiosPacific—that was willing to try to grow the cells in the test tube, rather than in mice. But another obstacle stood in the team's way. Growing antibodies in test tubes typically requires the use of fetal calf serum as a growth promoter. Calf serum is a gruesome byproduct of the slaughterhouse industry, and it has been hotly controversial, not only for the cruelty involved in its production, but also for its possible contamination with mad cow prions or other disease carriers. Fetal serum is also as biologically variable and unstandardized as it is cruel. So the PCRM team asked BiosPacific to work out a system of cellular growth promotion that sidestepped fetal calf serum . . . It took months of work, and the process was not cheap. But it eventually became clear that, indeed, the cells grew perfectly well with this method and produced the antibodies the team needed. The next challenge was to incorporate these antibodies into a test kit. To do that, PCRM worked with Linco Research of St. Charles, Missouri, one of the leading suppliers of insulin kits. It was several months before the kit was ready. Finally, several human blood samples were tested using the new system, and the results were compared to the existing insulin assays. When the data came in, the laboratory called PCRM to tell them the news: The new test was every bit as accurate as the old one—or perhaps just slightly better. [34]

Are there other examples where animal components have been replaced?

Researchers at the University of Wisconsin-Madison recently succeeded in growing human embryonic stem (HES) cells without using animal components. An article in *Nature Biotechnology* states the following:

> Human ES cell lines derived in defined conditions would be more directly applicable to clinical use than are cell lines derived in the presence of animal

> products. All of the human ES cell lines currently approved for federal funding in the United States were derived on mouse feeder layers and were exposed to a variety of other poorly defined animal products (see http://stemcells.nih.gov/research/registry). Derivation and culture in serum-free, animal product–free, feeder-independent conditions mean that new human ES cell lines could be qualitatively different from the original lines, and makes current public policy in the United States increasingly unsound. [35]

This breakthrough—growing human embryonic stem cells without using animal components—opens the door for stem cells to be used to treat patients suffering from diseases like Parkinson's, as the threat of disease and other complications from mixing animal components with HES cells will no longer exist. It should also bring to the fore the question of why we use animal products in the first place.

Are these examples of how animal components have been replaced with human- and synthetic-derived ones what's known as "alternatives"?

Yes. The word *alternative* when used *in this sense* refers to the replacement of a viable animal-based modality with a viable nonanimal-based modality. Because the word *alternative* here implies viability—it comes from the Latin *alternare*, meaning to interchange—in order for a modality to be an alternative, what it replaces must be scientifically viable in the first place. In this sense, a scientifically invalid practice cannot be replaced with a scientifically valid alternative.

What are the reasons why scientists would seek a nonanimal-based alternative?

The most obvious reason would be, as we have seen in the examples we discussed earlier, better science and better results. In addition, nonanimal-based methods may offer cost savings and other efficiencies.

Overall though, the search for alternatives is fueled by ethical concerns. While this book is not intended to address the ethical controversy of using animals in science, the question of alternatives touches on this issue.

Animal-based studies, insofar as they promise great value to human health and well-being, receive widespread support. In a moral cost-benefit analysis, the suffering of animals in biomedical research is seen as being more than offset by the consequent benefits to humans. The consensus in society is that using animals is a *necessary evil* in order to save humans.

However, if scientists conducting basic science research are not making claims for future application and are justifying their research grants simply on the grounds of obtaining more knowledge for the sake of knowledge alone, they are then faced with the problem of explaining to society their desire to use sentient animals for no reason other than the advancement of knowledge. Society has not historically been comfortable with this position. It is less clear what follows from this by way of public policy.

The alternatives to using ethically controversial animals, such as dogs and monkeys, include using such ethically noncontroversial animals as flies (like *Drosophila*) and worms (like *C. elegans*). It also includes using nonanimals, such as the fungus *Saccharomyces cerevisiae*, the bacteria *E. coli*, viruses, and plants. Indeed, less complex organisms are good models for pathway and gene analysis, because their core processes are likely to be preserved through evolutionary time. Human tissue is also a good resource for basic science research.

What about the role of technology in basic science research?

That is a very pertinent question because for decades, basic research in the biomedical sciences has been almost synonymous with animal-based studies. In reality, however, much of the necessary basic knowledge that has allowed medical advances has come from studying the fundamentals of chemistry, engineering, and physics. For example, advances like MRI scanners, the electrocardiogram, the electroencephalogram, PET scanners, ultrasound, radioimmunoassay, electron microscopes, and X-ray crystallography came from basic research in physics. The role of technology in bringing new insights into biological processes should become even more important as the technology itself becomes more and more sophisticated. However, at the present time, nonanimal basic research is seriously underfunded, which clearly represents a missed opportunity.

Are there other examples where the use of technology has provided an alternative to using animals?

Yes. Technology has greatly impacted the use of animals in the training of medical students. Medical simulators—where doctors-in-training are placed in a virtual environment that mimics the experience of performing an actual medical procedure—are enabling students to develop their skills without using animals. The simulators enable students to practice procedures over and over again without having to use a new animal each time. From the American College of Cardiology:

> Cardiologists can learn to perform risky catheter procedures such as carotid angiography on a virtual patient simulator, rather than on real patients, according to a new study in the May 2, 2006, issue of the *Journal of the American College of Cardiology*. "Virtual reality simulation technology has advanced to the point where we can actually use a virtual environment and have the trainee learn in a very 'patient-safe' way in a virtual patient environment and make mistakes on a virtual patient versus doing it on a real patient," said Christopher U. Cates, M.D., F.A.C.C., F.S.C.A.I. from the Emory University School of Medicine in Atlanta, Georgia . . . Previously, practitioners learning new catheter procedures practiced on animals, cadavers or mechanical models and then were supervised as they worked on their first live patients. The researchers are currently doing studies to see if the patients of practitioners trained on this simulator have better clinical outcomes. But the researchers say one advantage of simulator training is already apparent. The progress of trainees (their "learning curve") is tracked objectively, so evaluators don't have to rely on the subjective reports of an instructor . . . Dr. Cates predicted that simulator training will become as routine in medicine as it already is in the airline industry and other fields [36] . . . .

According to *Nature*:

> Simulation has developed hugely over the past decade. "It is a lot more than a couple of mannequins," says Bruce Jarrell, vice dean of research at the University of Maryland School of Medicine in Baltimore, which a little over a year ago opened its surgery simulation and technology centre. Students practise using surgical instruments to lift coils of rope viewed over a monitor, much as intestines are lifted during bowel surgery. They use the controls during a simulated endoscopy while watching a realistic duodenum on a monitor. Nurses learn to intubate a mannequin that can be programmed to respond to administered 'drugs' with changes in heart rate and blood pressure. And minimally invasive surgery is tried by students using instruments that mimic those used in actual surgery to clip an 'artery' — complete with 'blood' — during a simulated gall-bladder removal, viewed on a computer screen. The most advanced simulators have 'haptic feedback', which provides students with the sensation that their instruments are touching real tissue . . . Jonathan Lissauer, a doctor who recently trained at Johns Hopkins [one of the few remaining schools to use live animals], concedes the argument for animal use in medical research and advanced surgical training. He says that sometimes they were used "as just a diversion for people who won't be using those skills at all. I think then you cross the territory from appropriate medical education to something worse than that," he says. "There was no medical utility in having pigs die so that people going into psychiatry could play around. "From a purely academic perspective," he adds, "I thought there were substantial differences between human tissues and pig tissues — a lot of textural differences — and that the practising wasn't overly useful because of that." [37]

In 1994, 77 of 125 medical schools in the U.S. used live animals to teach medical students basic principles of pharmacology and physiology. That number, as of this writing (2008), is down to about eight, due in large part to technological advancements as well as the efforts of the Physicians Committee for Responsible Medicine (www.pcrm.org). Today, not one of the top ten medical colleges in the U.S., including Harvard, Stanford, Yale, Duke, and Columbia universities, uses animals to train doctors.

CHAPTER 3. Evolution and the Use of Animals as Predictive Models for Humans

Why is a discussion of the Theory of Evolution pertinent to the topic of animal use in science?

Because the Theory of Evolution shows us why using animals as predictive models for human drug and disease response is a scientifically invalid modality while we can still use animals for other purposes in science. Evolution illuminates why it *appears* we can use animals as predictive models—just like it appears that the sun orbits around the earth—but why in reality we cannot. Because the Theory of Evolution is the single organizing principle of modern biology, it enables us to put the facts of biology in their proper places and gives us the overarching theoretical framework for understanding the how's and why's of the remarkable similarities—and at the same time the profound differences—of all living things on earth.

But some people say that the Theory of Evolution is just that—a theory that's never been proven.

Those who say that the Theory of Evolution is *just* a theory do not understand that the word *theory* has a much different meaning in the world of science than it does in everyday speech.

In casual conversation, a person who says, "I have a theory..." generally means they are hazarding a guess about something. Often we use it to imply a lack of credibility, such as when someone says, "In *theory*, this is supposed to happen, but I'm not counting on it!"

In the world of science, though, a theory is an explanation of a set of observations. It is the end result of the scientific method that begins with observation, the formation of hypotheses to explain these observations, the subsequent testing of these hypotheses, and, if the hypotheses receive confirmation, subsequent replication of results. Theories in this sense are generally accepted to be true by the scientific community because they are supported by a lot of high quality evidence. Theories are also used to make predictions of events and to advance technology.

Why don't scientists just call it the Law of Evolution, like the Second Law of Thermodynamics?

There is actually little difference between theories and laws. Both begin as hypotheses, both subsequently turn out to be supported by vast amounts of high quality evidence, and both may be revised in the light of new, refuting evidence. Even Newton's Law of Gravity had to be abandoned in the light of evidence concerning peculiarities in the orbit of Mercury. (Newton's law of gravity is accurate enough for many practical purposes, e.g., baseball. To cover baseball *and* the orbit of Mercury, you really need Einstein's General Theory of Relativity.)

So what is the Theory of Evolution?

The Theory of Evolution is a set of interrelated ideas based on the principle that all living things evolved from simple organisms and changed through the ages by means of natural selection and numerous other mechanisms to produce millions of species. Evolution, as Darwin wrote about it, concerns populations of organisms, and the ways in which heritable traits displayed by such populations, such as wing shape in birds or the shape of a bird's beak, change over successive generations.

Like the word *theory,* the word *evolution* has other meanings. For instance, it can be used to mean *change over time* as when societies, companies, and individuals change over time we say *they evolved into what they are today* even though no heritable traits were involved. When Darwin was writing he was very familiar with the nautical meaning of evolution because it was used to mean *formations at sea.* Today, you will hear navy personnel speaking of *performing evolutions* when they mean they are practicing formation changes.

How do we know what biological evolution is?

In 1859, Charles Darwin incorporated his personal observations with the geological theory of the British scientist Sir Charles Lyell and the population theory of the British economist Thomas Robert Malthus to develop a theory of evolution. (Alfred Wallace did essentially the same thing at the same time.) Since then, scientists have gathered a huge body of evidence that documents evolution. This evidence has historically come from a number of sources, including fossils, biogeography (the study of the geographic distribution of species), embryology, and the study of vestigial organs, which are a species' useless remnants of organs that were once useful in its ancestral species. Evolution has also been documented through the direct observation of rapidly-evolving species.

What insights did Darwin derive from his observations?

Darwin's observations led to the theory that all living things on earth evolved from a single form of life that inhabited the planet at least 3.5 billion years ago. Over millions and millions of years, this single life form developed into multiple species through a branching process known as speciation. So, the organisms we see in the world today—including ourselves—are members of lineages that have descended from common ancestors in the distant past.

Because all species share a common heritage, which can be traced back to a single life form, all species share certain common characteristics. Any two species will have a common ancestor, and closely related species have a more recent common ancestor. The lineage leading to modern humans diverged from the one leading to mice around 70 million years ago. But the common ancestor of humans and chimpanzees lived far more recently—between 4 million and 10 million years ago.

Darwin's other insight concerned heritable variation in the clusters of animal populations that constitute a given species. Naturally occurring variation guarantees that there will be differences, not just between given members of given species, but between members of different species. After divergence from common ancestors, *within* species variation (also known as intraspecies variation) will, under the right circumstances, be gradually amplified to yield *between* species variation (also known as interspecies variation). Thus, evolution involves descent from common ancestors with subsequent modification as distinct lineages take different evolutionary trajectories.

What else did Darwin observe?

Darwin also figured out an important mechanism by means of which evolutionary changes can occur, which is known as natural selection. Throughout the animal world, there is an ongoing struggle for survival and reproduction. The physical environment poses significant challenges, as does the living environment, which includes predators, pathogens, parasites, and prey.

In a population of animals displaying heritable variation, some animals will be better positioned than others to meet these environmental challenges. These animals will be more likely to survive and reproduce at the expense of less well-favored ones. The offspring of the successful animals will likely inherit the characteristics that were advantageous to their ancestors. Their numbers will increase, and the result will be animal populations that change over successive generations.

It is important to understand that it is whole populations—not individual members of populations—that evolve over successive generations. Individual members do not evolve at all in the biological sense of the word.

When Darwin put forward his theory of evolution, he observed that the characteristics of organisms might change during the process of being passed on to offspring. However, because the principles of genetics were not yet known, he could not explain how or why these changes occurred. More recently, the advent of modern genetics and molecular biology has informed our understanding of how evolution operates at the molecular level.

How does evolution operate at the molecular level?

To explain that, we must understand a bit about the basic unit of life; the cell. All living things are made up of cells, and each of these cells has its own special function. For example, nerve cells carry messages to the brain, muscle cells control body movement, and so forth.

Figure 3.1

Deoxyribonucleic Acid (DNA)

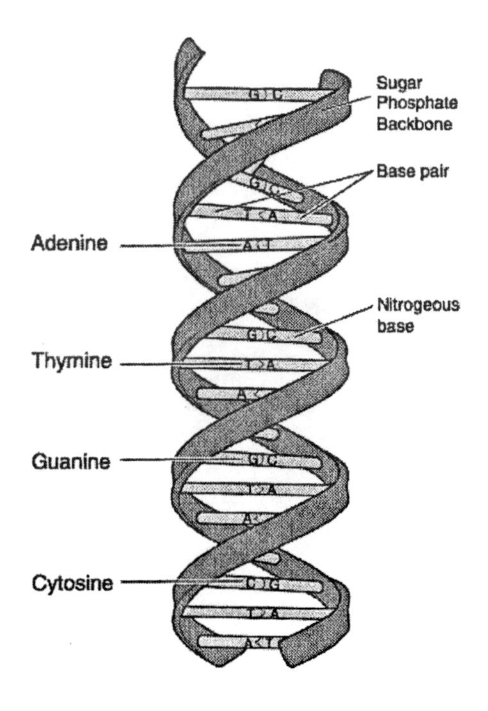

National
Institutes
of Health

National Human Genome Research Institute

Division of Intramural Research

The structure of a cell is made up chiefly of proteins, which in turn are made up of substances called *amino acids*. Proteins are critically important in building, maintaining, and repairing tissues in the body.

Just as each cell has a particular function, so do proteins. Enzymes are a type of protein that speeds up chemical reactions. Albumin, a protein in the blood, helps maintain the body's fluid balance by keeping fluids in the blood instead of allowing them to seep into the tissues. Antibodies are proteins in the blood that help protect the body from disease.

The nucleus is the control center of each cell—in other words, it directs all the cell's activities. Within the nucleus are the chromosomes. Chromosomes are long, threadlike strands made up of deoxyribonucleic acid (DNA), tightly coiled and packaged with proteins.

What is the structure and function of DNA?

DNA is the hereditary raw material of all organisms. It is composed of phosphate, a sugar called deoxyribose, and compounds called *bases*. There are four different bases: adenine (A), guanine (G), thymine (T), and cytosine (C). Adenine always pairs with thymine (AT), and guanine always pairs with cytosine (GC), to become *base pairs*. (See figure 3.1 on the previous page.)

Together, these components are arranged in chemically bonded units called *nucleotides*. The nucleotides bond to each other to form long chains called *polynucleotides*. A DNA molecule consists of two chains of polynucleotides arranged in a double helix, like a spiral staircase. The sides of the staircase are the polynucleotide chains of phosphates and sugars. The rungs are the base pairs.

What are genes?

Genes (see figure 3.3) are segments of DNA that encode the instructions that enable a cell to produce a protein. The *genome* contains the complete set of instructions for making the proteins that an organism needs to survive. *Gene expression* (see figure 3.4) is the process by which a cell makes a protein according to the instructions carried by the gene. Many genes form a chromosome (see figure 3.2).

What is involved in gene expression?

There are essentially two types of genes: structural and regulatory. Structural genes make the proteins our bodies need to make hands, kidneys, hearts, our immune system, and so forth. Regulatory genes tell the structural genes when and where to turn on and off, thus affecting the development of the embryo and the organism, as well as the physiology of the organism.

While every cell contains 100 percent of an organism's DNA, different cells need different proteins, so different genes are activated—that is, *expressed*—in different cells. The cell will activate whatever gene it needs; otherwise, the gene will be turned off. Some genes stay turned off for the entire life of an organism, while others are turned on almost continuously.

Figure 3.2

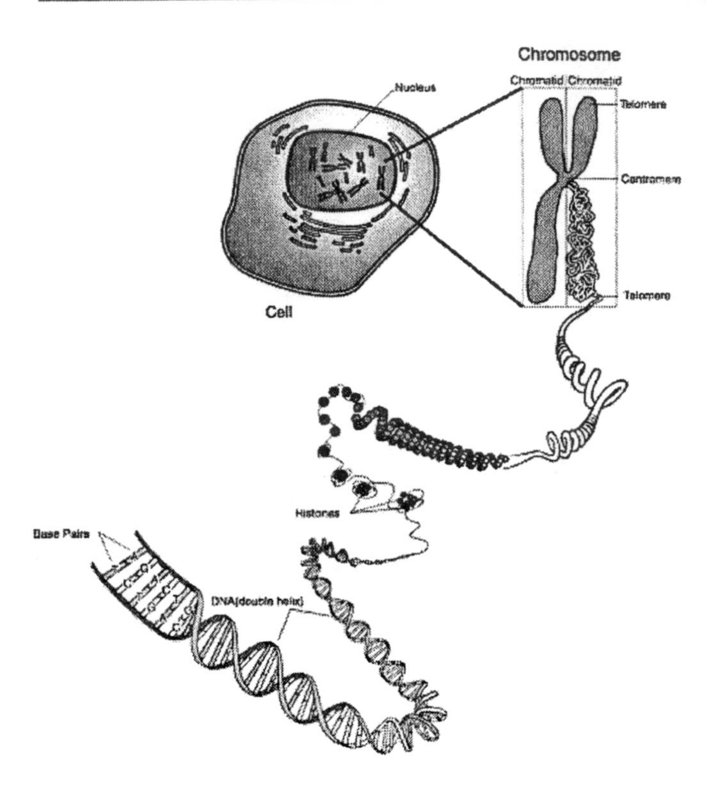

National Institutes of Health

National Human Genome Research Institute

Division of Intramural Research

Figure 3.3

Gene Advanced

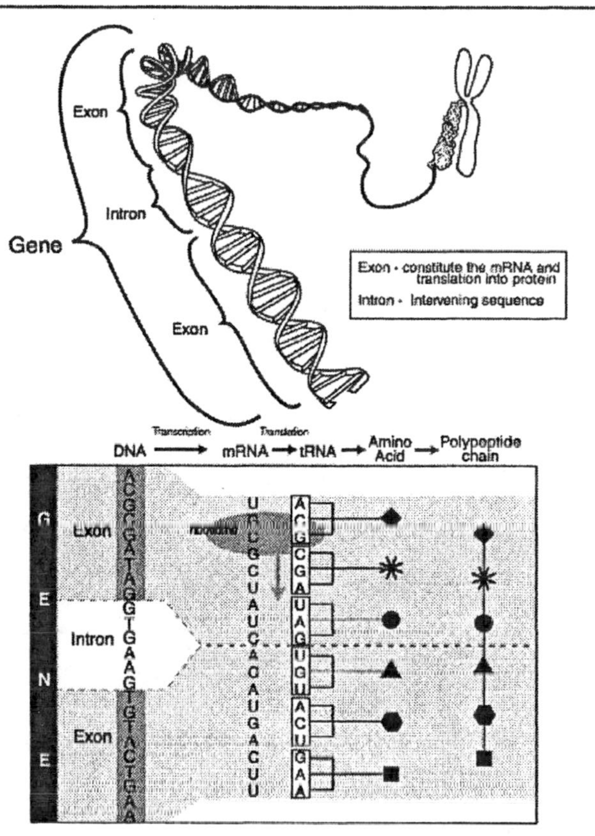

National Institutes of Health

National Human Genome Research Institute

Division of Intramural Research

When a gene is expressed, or switched on, that portion of the double helix unwinds, and the base pairs of AT and CG separate. That is, the G separates from the C, and the A from the T. The single strands are then "read" and copied to make ribonucleic acid, or RNA. RNA is composed of nucleotide bases like DNA. The RNA is then read and translated into amino acids, with each triplet of nucleotides coding for one amino acid. Amino acids combine to make proteins, which, you will recall, are essential to all plant and animal life.

Just as the order of letters in the alphabet make words, the order of the DNA base pairs—adenine/thymine and guanine/cytosine—determines the amino acid sequence, and the amino acid sequence determines which protein is made. Evolution occurs at the molecular level by the substitution of one nucleotide for another. A change in a single nucleotide can reorder the sequence of amino acids and hence make a different protein.

Does this process of gene expression occur in all organisms?

Yes. Whether you're talking about an elm tree, a monkey, yeast, or a human, life at its most basic, microscopic level is made up of the same DNA units, and they are assembled using the same process.

In fact, most animals actually have more or less the same genes. For example, roughly 93.5 percent of the macaque monkey's DNA base pairs are identical to humans. Chimpanzees are even closer—humans share 98.7 percent of their DNA.

Sean Carroll wrote in his excellent book *Endless Forms Most Beautiful*:

> The first shots in the Evo Devo [evolutionary and developmental biology] revolution revealed that despite their great differences in appearance and physiology, all complex animals—flies and flycatchers, dinosaurs and trilobites, butterflies and zebras and humans—share a common "tool kit" of "master" genes that govern the formation and patterning of their bodies and body parts . . . The important point to appreciate from the outset is that its discovery shattered our previous notions of animal relationships and of what made animals different, and opened up a whole new way of looking at evolution[27].

Then where are the differences? Obviously, humans are very different from macaque monkeys or mice.

By and large it is not the genes themselves, although some species and even individuals do have genes that others lack, but the activation of the genes combined with the variant forms they take in nature. Activate, or express, in one way and you get a human; activate another way and you get a mouse. Mice and humans both have the gene that allows mice to grow a tail. But, under normal circumstances the gene is not expressed in humans; hence, we have no tail. Carroll continues:

> We now know from sequencing the entire DNA of species (their genomes) that not only do flies and humans share a large cohort of developmental genes, but that mice and humans have virtually identical sets of about 29,000 genes . . .

Figure 3.4

Gene Expression

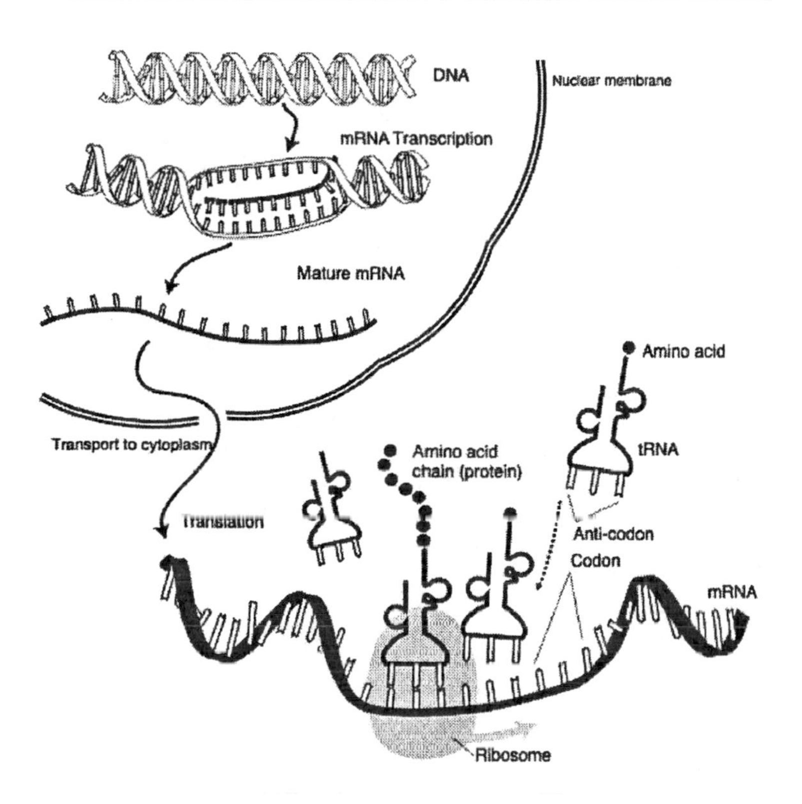

National
Institutes
of Health

National Human Genome Research Institute
Division of Intramural Research

> If the sets of genes are so widely shared, how do differences arise? . . . The first idea is that diversity is not so much a matter of the complement of genes in an animal's tool kit, but, in the words of Eric Clapton, "it's in the way that you use it." The development of form depends upon the turning on and off of genes at different times and places in the course of development. Differences in form arise from evolutionary changes in where and when genes are used, especially those genes that affect the number, shape, or size of a structure. We will see that there are many ways to change how genes are used and that this has created tremendous variety in body designs and the patterning of individual structures.

While most animals actually have more or less the same genes, what matters are not the genes *per se*, but the variant forms of a given gene that are often found with different statistical frequencies in different populations. Important differences between species also lie in when and for how long the genes are activated or turned on. You could probably use the exact same set of genes to make almost every mammal on earth by simply activating them at different times and for different lengths of time during the development and growth of an embryo.

It may help to think of the common genes involved in developmental processes as keys on a piano. The fingers that work the keys can be thought of as analogs of what geneticists refer to as *upstream regulators*. The strings hit by the activated hammers can then be thought of as downstream targets (themselves often regulatory in nature). All pianos have the same keys, but the music that is played depends on the person sitting at the keyboard. Starting with the middle C, we can play do-ra-me-fa-so-la-te-do (CDEFGABC). Depending on how the keys (genes) are played (expressed), we can play Chopsticks, Bach, Beethoven, Billy Joel, or Ray Charles. In fact, we can play an almost unlimited number of tunes on the piano using just the 88 keys. As Carroll states:

> The second idea concerns where in the genome is the "smoking gun." Around 3 percent [of our DNA] . . . is regulatory. This DNA determines when, where, and how much of a gene's product made . . .
>
> The whole tool kit of an animal contains several hundred or so different DNA-binding proteins, most with different signature preferences. There are an astronomical number of potential combinations of signature sequences in switches. If we assume a tool kit of 500 DNA-binding proteins in an animal, there are 500 x 500 = 250,000 different pairs of combinations of sequences and tool kit proteins. There are [25,000]00 x 500 = 12,500,000 different three-way combinations and over 6 billion different four-way combinations. These calculations illustrate the power of combinatorial logic of the tool kit and genetic switches . . . .

This simple example illustrates well why different individuals even of the same species respond differently to the same stimuli, say a drug. You may metabolize a drug rapidly while another person metabolizes it slowly; therefore, the safe and/or effective dose will be different for each of you. You may be allergic to penicillin, even though your mother is not. You may be susceptible to AIDS or lung cancer while your cousin is not. Even monozygotic (formerly called identical) twins suffer from different diseases.

How can monozygotic twins suffer from different diseases when they have the same genes? Aren't they exact copies of each other?

Yes and no. First, let's look at how monozygotic twins occur. In the process of fertilization, the male sex cell and the female sex cell are united, creating a zygote. The zygote is a single cell that contains the genes from both parents, and it develops into the embryo following the instruction encoded in its DNA. If the zygote, or one of the cells from the zygote, divides into another embryo, then we have monozygotic twins—that is, twins from a single zygote.

When sperm fertilizes two different eggs, it results in two different embryos, thus creating dizygotic twins (fraternal twins). Unlike monozygotic twins, dizygotic twins have unique DNA from the onset of formation. Monozygotic twins have identical DNA up until the time they become two separate embryos. That's when the interesting changes begin.

Because of differences in the womb (developmental noise) and differences in environments encountered after birth, monozygotic twins may display different features. (Geneticists discuss this phenomenon under the headings of *norms of reaction* and *phenotypic plasticity.*) In addition, the genes they do have in common may be expressed differently, leading to very different drug reactions and susceptibilities to disease. This is one reason why one monozygotic twin may suffer from a disease like multiple sclerosis, while the other does not.

What are gene networks?

In order to build a pancreas or a heart, many different genes must act together in concert. This coordinated action of genes occurs in the context of a gene network. Biologists are currently very interested in uncovering the structure or architecture of such networks. Gene networks are examples of *complex systems*. A complex system is one that is composed of many interconnected parts that feed back on one other, thus influencing each other.

The most important characteristic of a complex system, from our perspective, is that the whole is greater than the sum of its parts. Reductionism (breaking a system down into its isolated parts) cannot completely explain a complex system as a whole. Even when you know everything about every part, you will not be able to understand the whole.

The fact that living systems, as well as parts of living systems such as tissues and gene networks, are complex systems means that very small changes in a small section of one system can result in disproportionately large changes in the system as a whole. Evolution is the process of making very small changes to complex systems, which can ultimately result in different species.

How do genes influence the development of disease?

The causes of most human diseases lie in the proteins the body makes, the regulation of these proteins, and the protein-protein interactions. All these activities are determined by the process of gene expression, which, as we have described above, is the process by which a cell makes a protein according to the instructions carried by a gene.

Genetic instructions are so complicated that mistakes—known as mutations—can occur, and some of them lead to disease. Imagine that a regulatory gene imparts its instructions to the structural gene as one might type up a recipe for baking a cake. If the recipe is supposed to call for 1 cup of sugar, but a typographical error makes it 14 cups of sugar, the recipe will surely be a failure. Now consider the statement: all men are created equal. What if you added the letters *nt* after are? Changing just two letters profoundly alters the meaning of the statement.

Today, more than 4,000 diseases are thought to be caused by hereditary mutations. However, not all genetic mutations cause disease; some mutations are "silent"—in other words, the change does not cause an abnormal protein to be produced, and therefore it is not noticeable. In other cases, genetic mutation may be so lethal that it causes a fetus to abort spontaneously. In still other instances, a mutation causes an abnormal protein to be produced.

If almost all diseases are caused by genes, and if genes are composed of DNA, and the chemical composition of DNA is the same for all living things on earth, then why can't animals such as mice serve as models for human drug and disease response?

Since mice and humans diverged from their common ancestor, their respective lineages have followed different evolutionary paths. With time, both species evolved into what we see today. But as they evolved, they changed. Moreover, those changes are not insignificant, especially when you try to predict how one will respond to a drug based on how the other did.

We now know that with systems as complex as human and nonhuman animals, even very small differences may not be of negligible consequence. Some genes cause disease in members of one species, but not in members of another. And members of some species can have a gene removed without consequence, while members of another species will die without it. (Nor should we forget that members of a given species themselves display variation.) To give but one example, Jeffrey L Noebels of Baylor Medical School has observed:

> Perhaps more surprising is the realization that within this first group of human epilepsy genes to be described, there is not one the role of which in neuronal synchronization had been previously implicated in experimental animal models of acute epileptiform seizures. [38, 39]

Notwithstanding their inability to predict human response, our common evolutionary history does imply that animals can tell us important things about humans. After all, we do share many common characteristics and traits. From animals we can learn that cells are common building blocks of tissues, the way blood circulates, that life consists primarily of carbon, oxygen, hydrogen, and nitrogen, what an immune system is, and so forth. As long as the questions concern basic heritable properties, we can and have learned things from using animals.

What about using primates, since they are our closest genetic cousins?

In this case, close is not good enough when it comes to prediction. The devil of the differences lies in the details. Take the macaque genome project as an example. It revealed about 200 genes that evolved differently after humans and macaques branched off millions of years ago. Additionally, humans and macaques both possess genes that cause disease in humans but not in macaques. One example is the gene responsible for phenylketonuria or PKU (a genetic disorder that is characterized by an inability of the body to use the essential amino acid phenylalanine). It is present in both species, but only causes disease in humans [40]. Hence the warnings printed on cans of diet soda.

Regardless of the percentage of DNA similarity, studying another species will not predict human response to disease and drugs. Differences in gene regulation and expression, along with gene variants, account for many of these differences. Humans, monkeys, chimpanzees, and gorillas all share a common ancestor, but they react very differently to diseases and drugs – and humans and chimpanzees, for example, are more closely related to each other than either is to rats or mice!

Alfred North Whitehead said: "Seek simplicity but distrust it." It would be very simple to say that since all animals are evolutionarily related, we should be able to use one species to predict drug response for another. But it would be inconsistent with reality.

Can't one learn a lot about the human genome and how it works by studying the genome of an animal?

Yes, if what you want to learn about are the very fundamental genes that make the body. Science has learned much about the way the body plan goes together by studying fruit flies. But as your examination becomes finer, for example seeking the mechanism by which HIV enters the cell, then the differences begin to outweigh the similarities. Today, the questions we're asking are not so simple. In fact, they're incredibly complex. Advances in technology enable us to study human disease at the genetic level. And that is precisely where species differentiation is most pronounced. Rather than studying what makes us similar, we're studying what makes us different.

Studying genes in animals like mice is exactly what scientists did in the past when looking for a gene's function. This approach had decidedly mixed results and today humans are used more often than not. By studying a large number of human genomes, scientists can compare and contrast and find disease-causing genes. The study of genetic variation within our own species can convey medically relevant data in ways in which animal studies cannot.

Again, like studying intact whole animals, studying genes from animals will allow scientists to learn many interesting facts, but if one wants human-directed knowledge that can be used to cure diseases then human-based genes need to be studied. Using animals like fruit flies to study the fundamentals of genetics works very well and that is how many genetics classes are taught. One should never lose sight, however, of the fact that humans are not fruit flies writ large. In a similar vein, chimpanzees are not humans in ape suits.

Don't genetically engineered animals solve that problem?

Genetically engineered animals are those that have been created with certain genetic characteristics to suit the specific purposes of the researcher.

Genetically modified organisms (GMOs) can be divided in various ways. *Transgenic animals* are genetically engineered animals that carry genes from another species. *Knockouts* are genetically engineered animals that have had one or more of their genes removed to create a specific defect. Since some GMOs have had human genes inserted into their genetic code (*Knock-ins*), they are perceived to mimic more accurately the effects of human disease.

Given what you've learned about complex systems, gene expression, and gene networks in this chapter, you probably can see the problem with using genetically engineered animals to predict human response:

1. Complex systems are *robust*, which means that they are resistant to change. Inserting a gene or taking one out may be lethal or have no effect at all, depending on the individual.
2. Genes exist in copies, so removing one copy will not necessarily remove all the genes that serve the same function.
3. Because of what is called redundancy, different genes can combine to make the same protein that the missing gene made.
4. Adding a gene does not mean it will be expressed.
5. Just because a gene causes a certain thing to happen to a dog or mouse does not mean that the same thing will happen in a human. It does not even mean it will result in the same outcome in a different mouse. A study in *Science* [41, 42] revealed that one strain of mice could have a gene removed without any real consequence, while another strain would die without the gene.
6. Rarely does it occur in nature that there is a one-to-one correlation to genes and the effect they have on an organism. Although there are diseases that are the result of a single damaged gene in one's genetic code, one cannot assume one gene to one result. In the vast majority of circumstances, it's just not that simple.
7. Remember, genes act as part of a network rather than a single unit. One gene may interact with others to create a specific result. And several functions may be associated with a single gene. Inserting a damaged gene into the genome of an animal, for example, may produce activity that may be unrelated to the goals of the experiment, and which cannot be predicted.

Genetically engineered animals are certainly interesting and can be useful in science if the right questions are asked. But using transgenic animals to predict human response to drugs and disease does not make sense from the perspective of evolutionary biology.

Chapter 4. Prediction and Animals

Why do you say that animals have failed as a valid modality for predicting human drug and disease response?

As the previous chapter has shown, the Theory of Evolution provides us a theoretical framework while advancements in genetic and molecular biology provide the empirical data as to why animals cannot adequately predict human drug and disease response. Mice are unique and interesting creatures in their own right, they are not simply men writ small! In this and the next chapter, we will see how the animal model has failed at the clinical level, particularly in the search for treatments and cures for some of the major diseases.

Animals and humans share many similarities in terms of the "stuff" they are made from (all have cells, genes, lipids, proteins, and so on), but they also exhibit many differences. At the subcellular and genetic level, where the vast majority of research is now taking place, *organizational differences* between animals and humans outweigh the similarities in ways that are relevant to a discussion of prediction. (Recognition of organizational differences draws our attention to the way "stuff" is put together, used and regulated).

However, it remains true that some of those who insist that animals can be *predictive* models do not understand the use of the word *prediction*; in this chapter, we will explore what prediction means in the world of science, and how animals fail to meet a proper scientific standard of prediction.

What is the scientific definition of the word prediction?

The way scientists use the word *prediction* is different from the nonscientific use. Prediction, like science itself, is difficult to define in a single sentence. It is easier to describe it then define it but descriptions and definitions usually include the following attributes:

1. A particular outcome will occur.
2. The outcome that will occur will occur in the future.
3. It can refer to a "correct statement about some presently unknown phenomenon, whether it is a past, present, or future event. Examples include predicting that the extinction of dinosaurs was caused by a collision on earth of an enormous meteor [43]."

4. If the outcome is not in the future, then an analysis of the past does not involve cherry picking the data.
5. As new knowledge emerges, new predictions should be possible.
6. The outcome of the prediction is very specific.
7. The conditions under which the outcome will occur are very specific.
8. The outcome can be expressed qualitatively and quantitatively if possible. For example, the sensitivity, specificity, positive predictive value, and negative predictive value of tests should be known.
9. Conditions exist which would falsify the prediction.

These definitions are a far cry from the everyday usage of the word prediction. Consider this example: A fortune teller says you will get married this year, have a baby next year, make $100,000 the next, discover a cure for cancer the next, and retire the next year. Three years after the fortuneteller read your palm, you did indeed make $100,000. Did the fortuneteller predict this? In lay terms, you could certainly say so. But science is very specific, and in this case, the fortuneteller's declarations do not meet the scientific standard for prediction as outlined above.

Let's take another example. A university has a history department comprised of 200 faculty and staff. This university also has a football team, and hence the campus is very interested in football. At the beginning of the season, the 200 people in the history department wager on who will be the ultimate #1 team at the end of the season. Considering the finite number of serious contenders, it is highly probable at least one of the members of the history department will pick the winner. Can we then say that the history department always predicts the winner? No, for two reasons.

First, if we are going to say the history department *per se* picked the winner, then the history department would get *one* pick, not 200. Second, prediction in the scientific sense does not mean merely a correlation of outcomes. If we are to say that the history department is capable of *predicting* (in the scientific sense of the word), then we would need to see evidence that out of the last 20 or so football champions (whatever the actual number we need a reasonable track record of success), the history department picked the winner a very large percentage of the time. Getting it right just once could be a fluke. One right guess does not a prediction make, at least not in science.

Why does science make predictions?

According to Salmon in *Philosophy of Science*, there are at least three reasons for making predictions:

1. Because we want to know what will happen in the future;
2. Because we want to test a theory; and
3. An action is required and the best way to choose which action is to predict the future. [44]

For example, say we are testing a chemical for carcinogenesis—that is, we want to know if that particular chemical will cause cancer if ingested. So we apply #1: what will happen in the future, and #3: an action is required. In this case, the action is whether or not to allow the chemical on the market, and the best way is to choose which action will predict the future. Neither #1 nor #3 is subtle. We want a correct answer to the question, "Is this chemical carcinogenic to humans?" That is why we test drugs and chemicals before releasing them—to predict what they will do.

In science, how is prediction different from guessing correctly and finding correlations?

In science, guessing correctly or finding correlations are not the same as predicting the answer. A fundamental part of any theory or practice claiming predictability is its ability to predict the result of an experiment that has not yet been done. For example, the Second Law of Thermodynamics predicts that a perpetual motion machine can never exist, and that there are upper limits to the efficiency of engines, such as the one in your car. The field of astronomy uses the laws of physics to predict the location of planets.

Again: In science, if you claim predictability, you must be able to predict the result of an experiment that has not yet been done. This concept separates the scientific use of the word *predict* from the lay use of the word, which more closely resembles words such as guess and conjecture—as we saw with the fortune teller and the university history department.

Where does the burden of proof lie in terms of prediction?

Those who claim that animal models are predictive must demonstrate that this claim is correct. The evidential burden of proof resides with those who make the claims. Do you believe that Big Foot exists and is roaming the forests of the Pacific Northwest? Then the burden of proof is on you to prove to us skeptics that Big Foot does exist. It is no use saying to the skeptic, "You haven't shown me that Big Foot isn't there."

Neither is it our responsibility to provide examples of scientifically viable and predictive tests that do predict carcinogenicity. A modality is either predictive or it is not, regardless of what else is available. Astrology is not a predictive modality for learning about future marriage prospects. Neither is anything else to the best of our knowledge, but that does not mean that astrology wins by default.

To paraphrase a popular adage, "show me the data." If someone suggests that an animal—say, a mouse—can predict human response to chemicals vis-à-vis carcinogenesis, she would need to provide data that support her claim. Perhaps no one animal alone is capable of predicting human responses. But when the same result occurs in two species—say a mouse and a monkey—then perhaps the results are predictive. But the need to provide the appropriate statistical data to prove that claim is the same.

What are the criteria for determining the predictive value of an animal model—in other words, what kind of statistical data would prove that an animal model is or isn't predictive?

In biology, many concepts are best evaluated by using simple statistics like the four described below. By using these four measures we can ascertain whether a test does what we want it to do—whether it is predictive. The criterion for predictability lies in how the test measures up in these four areas.

For example, in medicine we can use a blood test to determine whether someone has liver disease. In order to ascertain how well this test actually determines the health of the liver, we calculate the following:

1. The sensitivity of the test;
2. The specificity of the test;
3. The positive predictive value (PPV); and
4. The negative predictive value (NPV).

The *sensitivity* of a test is the probability of a positive test among people whose test should be positive. In this case, the sensitivity of the blood test in our example would be the probability of a positive test among people who do have liver disease.

The *specificity* of a test is the probability of a negative test among people whose test should be negative. In our example, the specificity of the test would measure the probability of a negative test among people who do not have liver disease.

The *positive predictive value* (PPV) of a test is the proportion of people with positive test results who are actually positive. In our example, the PPV of the test would measure the proportion of people with positive test results who actually have liver disease.

The *negative predictive value* (NPV) is the proportion of people with negative test results who are actually negative. In our example, the NPV of the test would measure the proportion of people with a negative test who actually do not have liver disease.

All these values are measured on a scale from 0.0 (being the lowest) to 1.0 (being the highest). Very few tests have a sensitivity, specificity, PPV, or NPV of 1.0. But in order for the test to be useful—in this case, to tell us if the patient actually has liver disease—it needs to be predictive far more often than not.

How do these measurements apply to animal tests?

Before we answer that question you need to understand how science in general works vis-à-vis what counts as evidence. Scientists conduct experiments of various types and then submit their results and thoughts to journals that have referees. Referees are scientists in the same field (*peers* of the scientist submitting the paper) who judge whether the experiments, results, and the scientists' thoughts about it all are worthy of being published. There are many flaws to this system but all in all it works. The question we are considering, whether animals are predictive for humans, would be answered in part by scientists comparing the results of animal tests with the results from humans taking the drug or suffering from the disease. These results would be subjected to the four tests described in the previous question. Then the resulting paper would be submitted and hopefully accepted to a scientific journal and the experiment and results made available to the scientific community. This entire process is sometimes referred to as *peer review* and the scientific journals are called *peer reviewed journals*.

We can show this best by a specific examples. The data from testing six drugs on animals were compared with the data from humans and published in a peer review journal [45]. The animal tests were shown to have a sensitivity of 0.52 and the PPV was 0.31. The sensitivity is about what one would expect from a coin toss, and the PPV even less. This is not considered predictive in the scientific sense of the word.

Two studies conducted in the 1990s and also published in peer reviewed journals were equally revealing. One showed that out of 24 human toxicities only 4 were found in animals [46]. In another study, in only 6 of 114 cases did clinical toxicities have animal correlates [47]. Fletcher reported on drug safety tests and subsequent clinical experience with 45 major new drugs. Some effects were seen only in animals, while others were observed only in man. The survey established that 25 percent of toxic effects observed in animals might be expected to occur as adverse reactions in man. [48] Lumley:

> In one small series in which the toxicity in clinical trials led to the termination of drug development, it was found that in 16/24 (67%) cases the toxicity was not predicted in animals. [49]

Many other studies have found the same lack of predictability.

The sensitivity, specificity, PPV, and NPV of animal models based on these studies are obviously suboptimal. Unless a test as a whole has a highly positive predictive value (PPV), the test is not useful. (See *Animal Models in Light of Evolution* for more such experiments.)

How do animal experimenters respond to data that refute their claim that animal models are predictive?

First, they cherry pick the data. We will discuss cherry picking in chapter 8 but briefly, cherry picking means surveying many results from animals species and choosing only the ones that show what you want them to show. They ignore all the results that disagree with their hypothesis.

Second, because the animal tests have failed the four criteria described above, sensitivity, specificity, positive and negative predictive values, researchers sometimes make up new specious tests that give the results they want. Studies that superficially look like the above examples have animal experimenters using the spurious statistics like *concordance rate* or *true positive concordance rate* when they judge the validity of animal tests. However, these terms are not in the normal prediction-relevant lexicon and are usually used to mean correlation, which has nothing to due with prediction.

For example, a study was performed in the late 1990s by Olson et al., [50] which introduced the term *true positive concordance rate* when what they were really talking about was simply sensitivity—something that does not come close to settling the prediction question. This has not stopped this misleading study (sponsored by the drug industry) for being cited in support of predictive modeling claims). [See [5] for more on this.]

Sometimes, though, we don't even need to do the math. An irreverent aphorism in biology known as Morton's Law states: "If rats are experimented upon, they will develop cancer." Morton's Law is similar to Karnofsky's law in teratology (the study of birth defects), which states that any compound can be teratogenic—that is, cause birth defects—if given to the right species at the right dosage at the right time in the pregnancy. The point here is that it is very easy to find positive results for carcinogenicity and teratogenicity, but this is meaningless without also knowing how many times the animal got it wrong.

It is true that all known human carcinogens that have been adequately studied have been shown to be carcinogenic in at least one animal species [51-53]. However, prediction does not mean *retrospectively*—that is, after the fact—finding one animal that responded to stimuli like humans and then saying that the animal *predicted* human response.

Even if science did decide to abandon the historically correct use of the word *predict*—the chances of which are slim to none—every time an animal-model advocate said that animal species X predicted human response Y, she would also have to admit that animal species A, B, C, D, E, and so on failed to be predictive. The relevant questions are: How are we to know *prospectively*—that is, before the fact—which animal will mimic humans? And what would be the value of such a prediction?

The thalidomide tragedy of the 1950s is a good example of the issues raised by predictability, prospective knowledge, and retrospective knowledge.

What happened with thalidomide in the 1950s?

Thalidomide was a drug released in the late 1950s as a treatment for morning sickness in pregnant women. Unfortunately, the babies of the women who took the drug were born with phocomelia (the absence or deformation of limbs) as well as other anomalies. At least 10,000 children were affected.

Is it true, as advocates for animal experimentation say, that more animal testing would have prevented this tragedy?

Schardein points out:

> In approximately 10 strains of rats, 15 strains of mice, 11 breeds of rabbits, 2 breeds of dogs, 3 strains of hamsters, 8 species of primates and in other such varied species as cats, armadillos, guinea pigs, swine and ferrets in which thalidomide has been tested, teratogenic effects have been induced only occasionally. [54] [p5]

It's important to note that the above tests were performed *after* the thalidomide disaster. But the question we are asking is whether animal testing *per se*, if performed appropriately, *would* have prevented the disaster. So examining the results from all possible tests is acceptable. Schardein again:

> It is the actual results of teratogenicity testing in primates which have been most disappointing in consideration of these animals' possible use as a predictive model. While some nine subhuman primates (all but the bush baby) have demonstrated the characteristic limb defects observed in humans when administered thalidomide, the results with 83 other agents with which primates have been tested are less than perfect. Of the 15 listed putative human

> teratogens tested in nonhuman primates, only eight were also teratogenic in one or more of the various species. . . .[55]

Manson and Wise summarized the thalidomide testing as follows:

> An unexpected finding was that the mouse and rat were resistant, the rabbit and hamster variably responsive, and certain strains of primates were sensitive to thalidomide developmental toxicity. Different strains of the same species of animals were also found to have highly variable sensitivity to thalidomide. Factors such as differences in absorption, distribution, biotransformation, and placental transfer have been ruled out as causes of the variability in species and strain sensitivity[56].

Would extensive use of the animal model have predicted thalidomide's adverse affects? No. Not a single individual species or combination of species can be said to have predicted the effects of thalidomide. If we were going to say the animal model *per se* predicted thalidomide's effect on pregnant humans, then the animal model would get one prediction. That one prediction could be based on one animal or a combination of animals, but there would ultimately be only one prediction.

Suppose a target is painted on the side of a barn. Even someone who had never fired a gun before will eventually hit the bull's eye if he fired enough bullets at the target. But you could hardly call this novice a marksman. The true marksman is the one who has one bullet, and hits the bull's eye on the first shot. The novice may also hit the target with a lucky first shot, but being a novice and not a marksman he will not often repeat this feat.

But some animals did react as humans. Wasn't that useful knowledge?

No. Since thalidomide was tested on a hundred or so species and strains, it is not surprising that some animals reacted as humans. But to determine if that knowledge would have indeed been useful, we need to return to the scientific use of the word prediction. Some scientists note that a specific breed of rabbit, known as the White New Zealand, suffered from the same birth defect, phocomelia, as the babies born to women who took thalidomide. So, did the White New Zealand rabbit *predict* phocomelia? In lay terms one could certainly say so. We all occasionally brag, "I predicted the outcome of the ball game/ election/ controversy." But remember, in science the word prediction has a very specific meaning, and the case of the White New Zealand rabbit does not qualify.

Prediction does not mean merely a correlation of outcomes. If we are to say that the White New Zealand rabbit is a legitimate *predictor*, then we would need to see evidence that a very large percentage of the time it reacts to teratogens (chemicals that cause birth defects) as humans do. In other words, we need evidence indicating a track record of predictive success with other substances (that it was sensitive and specific and had high positive and negative predictive values, and so on). Such a track record of success would at least give some reason to suppose that this strain of rabbit might be useful in the case of a new substance. There was no such track record at the time of testing, and alas, prior to actual testing, The White New Zealand rabbit did not come with a magical label saying, "Guaranteed good only for thalidomide testing." The White New Zealand rabbit was the exception, not the rule, when it came to thalidomide toxicity.

This is important because the animals that suffered phocomelia from thalidomide were found on an *ex post facto* basis *after* the drug's effects on humans were known. Useful predictions in toxicology need to be made *prior to* the occurrence of the events at issue. Retrospective correlation is *not* prediction (at least if safety is the issue). Add to the above the fact that all the animals that exhibited phocomelia did so only after being given doses 25-150 times the human dose [57-59]. It does not appear that any animal or the animal model *per se* predicted thalidomide's teratogenicity in humans, even after the fact of the human tragedy.

What would happen if we subjected thalidomide to the same testing regimen that is required today by the federal government?

Let's first go back to Karnofsky's law in teratology, which states that any compound can be teratogenic if given to the right species at the right dosage at the right time in the pregnancy. In all likelihood, if any drug currently on the market were subjected to the same testing regime as thalidomide, we would likely see some species suffer birth defects. Given thalidomide's profile today, physicians would advise pregnant women not to take the drug, which is what physicians advise every pregnant woman about almost every nonlife-saving drug anyway.

If we did not allow on the market any chemical or drug (drugs are a type of chemical as are the ingredients in deodorants, soaps, shampoos and so forth) that causes cancer in any species, then we would have no chemicals or drugs at all. Furthermore, there is a cost to keeping otherwise good chemicals off the market. We lose treatments, and perhaps even cures, for disease. For example, thalidomide has reemerged as a treatment for dermatologic diseases. Keeping drugs off the market sacrifices the income that could have been generated by their sale. We also lose new knowledge that could have been gained from learning more about the chemical.

Sir Arthur Conan Doyle said via his character Sherlock Holmes: "There is nothing as deceptive as an obvious fact." Aldous Huxley said: "Facts don't cease to exist because they are ignored." Animals are not predictive for disease and drug response and that fact is not going to change just because it conflicts with the agenda of some.

Aren't animals still useful even if they only get it right occasionally?

If by useful you mean, "Can we gain knowledge from them?"—then the answer is certainly *yes*. An experiment will always result in knowledge—even if we only learn what doesn't work. But if you mean "Will the results from animals predict what a drug will do in humans or the natural progression of a disease—how the disease affects the body, what it does to the cell, what happens if left untreated?"—then the answer is *no*.

There is a difference between *useful* and *predictive*. Occasionally getting it right is not the same as prediction, and in the case of disease study and drug testing, animal models are not predictive. It is difficult to imagine how a modality could be useful if the purpose of the modality—prediction—was not realized. We are not the only ones who realize animal models are not predictive: Lin in 1995 compared drug parameters in different species:

> Although the validity of animal testing to predict efficacy and safety in humans has been questioned, it is generally believed that data from animal studies can be reasonably extrapolated to humans with the application of appropriate pharmacokinetic principles . . . In fact, most of the reported literature successfully showing the animal-human prediction, particularly those based on an allometric approach, are essentially hindsight and not predictive. [60]

Eason:

> The failure of animal toxicity studies to predict drug toxicity in humans, due to species differences in metabolism and pharmacokinetics, is illustrated by reference to the anti-inflammatory antiviral terpenoid carbenoxolone, the antiasthmatic candidate drug FPL 52757, and the cardiotonic drug amrinone. [61]

(See Chapter 6 for more scientists who agree animal models are not predictive.)

What about the times an animal and human reacted the same way to a drug? Wasn't the animal useful in that case and doesn't that justify their use?
No. As of 1980, there were roughly 1,600 known chemicals that cause cancer in mice and other rodents. But only about 15 of these chemicals caused cancer in humans[62]. What, prospectively, distinguished the 15 human carcinogens from all the others? Nothing. That is why animal tests are not even useful in terms of deciding what drugs make it to market, what the mechanism of a disease is and so forth. Only about 19 drugs have been shown to be teratogens in humans, while over 800 have been shown to be so in animals [54, 63-65]. What, prospectively, distinguished the 19 human teratogens from all the rest? Again, nothing.

Obviously, there will always be overlap between species. But since we can only know in retrospect which chemicals are harmful to both humans and animals, the overlap is meaningless for helping us decide what drugs to put in the marketplace.

When we examine the results of animal testing, we are looking at disagreements among animals, and hence must make our decision based on the value of the test *per se*. Look at table 4.1 on the next page and decide if you would approve the drug based on the animal test result listed. Then compare your choice to the table 4.2 on the next page, which shows the effects on humans for the drugs listed.

Figure 4.1

Drug	Human	Animal species								
		Unknown	Rat	Cat	Dog	Mouse	Monkey	Guinea pig	Rabbit	Hamster
						Lung CA				
			Safe	Safe	Safe	Safe	Birth defects	Safe	Birth defects	Safe
				Lethal						
				Safe	Safe	Safe	Safe	Safe	Safe	
			Cancer (male)	Safe	Safe	Safe	Safe	Safe	Safe	
							Lethal			
					Safe	Safe				
				killed 1 cat	Safe	Safe	Safe	Lethal	Safe	Lethal
						Safe	Safe	Safe	Safe	
					Safe	Safe	Mixed			
					Safe	Safe	Safe			
					Safe	Safe				
			Hepatotoxicity			Hepatotoxicity				Hepatotoxicity
					Safe	Safe				
			Safe		Lethal		Safe		Safe	
						Mixed (Strain specific)				
		cardiotoxic								
			Safe	Safe	Safe	Safe	Safe	Safe	Safe	

Figure 4.2

Drug	Human	Animal species								
		Unknown	Rat	Cat	Dog	Mouse	Monkey	Guinea pig	Rabbit	Hamster
Isoniazid	Safe					Lung CA				
Thalidomide	AIDS, CA Birth defects		Safe	Safe	Safe	Safe	Birth defects	Safe	Birth defects	Safe
Acetaminophen	Safe			Lethal						
Fenclozic acid	Liver failure			Safe	Safe	Safe	Safe	Safe	Safe	
Saccharin	Safe		Cancer (male)	Safe	Safe	Safe	Safe	Safe	Safe	
Actinomycin-D							Lethal			
Flosint	DEATH				Safe	Safe				
Penicillin	Cure			killed 1 cat	Safe	Safe	Safe	Lethal	Safe	Lethal
AN-1792 (AlzheimerÕs)	Worsens disease					Safe	Safe	Safe	Safe	
Vioxx	Heart attack Strokes				Safe	Safe	Mixed			
HRT	Cancer, Dementia, Heart disease				Safe	Safe	Safe			
Fen-phen	Heart valve abnormalities				Safe	Safe				
Furosemide	Safe		Hepatotoxicity			Hepatotoxicity				Hepatotoxicity
Opren	DEATH				Safe	Safe				
chocolate	Safe		Safe		Lethal		Safe		Safe	
chloroform	safe					Mixed (Strain specific)				
Digitalis		cardiotoxic								
Ipecac										
fenclozic acid	cholestatic jaundice		Safe	Safe	Safe	Safe	Safe	Safe	Safe	

Don't you need to test drugs and study diseases in whole intact mammals?
The only intact mammal that can adequately predict drug response in you is—you. The *intact systems argument*, to which this question refers, has been one of the mainstays of the animal experimentation community. This argument holds that we will not know how a disease is going to affect humans or how a drug will affect humans without studying an intact close evolutionary relative like a dog, mouse, or monkey.

The argument has merit in the sense that studying diseases and drugs in cell culture medium or on a computer will not give you the entire story of what will happen in humans. But the argument is wrong in its claim that studying intact whole mammals will give you the entire story of what will happen in humans because it assumes that animals are predictive, and they are not. The reason for this is that humans and the animals used to model them exhibit causally relevant organizational differences. Mice and men may be intact systems, but they are differently organized intact systems. The line leading to modern mice diverged from the line leading to modern humans about 70 million years ago – for 140 million years of independent evolution and adaptation to diverse challenges posed by nature. From the standpoint of evolutionary biology, it is frankly bizarre to suppose that having pursued such divergent evolutionary trajectories, mice are men writ small. (See *Animal Models in the Light of Evolution* for a more detailed discussion of these matters).

Have you been able to quantify the number and types of experiments on animals that failed to be predictive?
There are a few reasons why it is difficult in many instances to quantify the number and types of animal experiments that have failed to be predictive.

1. Negative data are not usually published.
2. Drug companies own their data, and they are proprietary; the companies are under no obligation to release the data to the public. Therefore, it is almost impossible to find all animal data on a drug in use today. Why would a drug company want you to see data that their drug harmed an animal when they are trying to sell it to you? At best, such data would merely reinforce the claim that they are relying on data that has no meaning to humans.
3. Very few review articles are written criticizing a practice that brings money to the institution that employs the author.
4. It takes a very long time to map out how the discovery of a particular drug occurred. Misleading animal data will be forgotten and only data that turned out to be representative of the ideal will be included. (See Peter Medawar's paper on fraud at: http://contanatura-hemeroteca.weblog.com.pt/arquivo/medawar_paper_fraud.pdf.)
5. When specific examples of the failure of the animal model are brought to the attention of animal experimenters, they are often dismissed as "an exception to the rule."

There is a definite paucity of data available on animal tests—a situation that animal experimenters try to use to their advantage. But when predictivity is defined properly and the results of animal-based studies are examined, it is clear that animals are not predictive for human drug and disease response.

If animals were involved in the discovery of a particular drug, disease, treatment, or biological process, doesn't it follow that they were necessary for that discovery?

This is a common claim that needs careful attention. Here is an example: Americans for Medical Progress, a pro-animal experiment industry group, lists on its website anesthesia, antibiotics, vaccines, vitamins, and allergy treatments as just some of the lifesaving treatments in which animal-based research was involved [66]. A link to another page provides an even longer list, with more details. They make it appear that without the use of animals none of these breakthroughs would have been possible, which is not true. One should always be careful to note the wording when encountering such assertions. Most of the time the claimant is saying animals were involved, not that animals were necessary. The two are not the same. If the claimant is saying that animals were necessary then he needs to support that claim with evidence and argumentation. Making remarkable claims is easy but proving them is usually difficult, so PR groups and others who do not have scientific truth as their primary interest avoid doing so. In fact, demonstrating that a medical science discovery could only have been made by using animals—that animals were necessary (not merely useful or important as a contingent, accidental fact of history)—is a very tricky proposition. Simply pointing to examples is not enough. The claimant must show that the discovery in question could not have been made any other way. This must be done for the claimant to say the discovery was necessarily dependent on animal use.

Do you have an example of such an instance?

Here is an example with a cautionary note. The discovery of the mechanism of the action potential (how the nerve transmits information) by Hodgkin and Huxley in the 1940s was made possible because they conducted experiments on the axon of squid, which have the largest nerves in the animal kingdom,. The large diameter of the squid's axon enabled Hodgkin and Huxley to insert voltage clamp electrodes inside the lumen of the axon. Supposedly, to this day no experimental preparation yields greater accuracy in the measurement of action potential characteristics, and thus squid are still widely used.

In order to prove Hodgkin and Huxley could not have made their discovery without using animals, however, we would need to know why smaller axons could not be used. Are there limits to the mechanical engineering of the instruments? Or, is it just the case that no one had tried to change things since this preparation was adequate? The burden of proof for stating that such use of squid is the *only* way these discoveries could have been made falls to the person making the claim. Good luck!

When evaluating claims, consider what the claimant is actually saying and compare it with how the statement is likely to be interpreted. This can give you some idea of the claimant's intent. We will explore in greater detail how to evaluate claims of the predictivity of the animal model in later chapters.

Didn't all winners of the Nobel Prize in medicine or physiology experiment on animals?

Probably, most did. But it is important to consider the following:

1. In no case does that mean the discoveries could not have been made without animals. It only means that animal-based studies were common and used by the researcher.
2. From the second half of the 19th century forward, experimentation on animals became part of all medical curricula. So researchers were usually obliged to perform animal experiments to get their degrees. But it is hardly accurate to deduce that those experiments bore directly on the Nobel-winning results.

Even if you are correct in claiming that animals are not predictive models, isn't there something to be said for the value they may have had in past discoveries?

Of course. It is true that animals have been used in many discoveries and enterprises for centuries. And it is equally true that some of those discoveries no doubt could have been made without animals, and some could not have been made without animals. Likewise, some could not have been made without using animals *at the time*, but could have been made later without animals due to advancements in technology.

The real point is that the history of medicine is complex and difficult to interpret. Careful detailed studies of historical episodes rarely support the *Saints and Heroes* stories that medical educators pass on to their young charges. Moreover, the questions that were being asked in the past are far different from those being asked today, and what is needed is an understanding of which modalities are best suited to obtain the answers. Society and scientists should distinguish between where animals can be viably used and where they cannot.

Doesn't the fact that thousands of drugs have been approved for human use, and that the vast majority of these are harmless, refute your argument that using animals as predictive models is a failed modality?

No, it doesn't. First, a vast majority of drugs have been approved for added uses or were me-too drugs, which are drugs that were very similar to others already in distribution. Many drugs were already known to be safe and were just reformulated to last longer, or they were approved for different uses that came to light after they had already been released in the market. These approvals are not new in the sense of animal models predicting their safety or efficacy.

Second, most drugs are not, in fact safe for everyone. Nor are they efficacious for everyone. Think of the numerous drug recalls reported in the news because of previously unknown side effects. The animal models did not predict a vast majority of these side effects. Furthermore, animals are not even used in an attempt to predict subjective reactions to drugs such as nausea, headaches, dizziness, and so forth—which also happen to be some of the most common side effects.

Third, the sad fact is that many people have died from drugs that tested safely in animals. An unnamed clinician quoted in *Science* stated, "If you were to look in [a big company's] files for testing small-molecule drugs [in humans] you'd find hundreds of deaths [67]."

Can you provide specific examples of instances where humans were harmed directly or indirectly as a result of relying on data from animal models?

There are far more examples than the limited space in this book can provide, but here are a few:

- Asbestos and other environmental toxins were thought safe because of animal testing.
- Animals did not suffer from atherosclerosis when fed a high cholesterol diet, so in the 1950s scientists said that cholesterol was unrelated to heart disease.
- Artificial heart valves were delayed because dogs react to the valves differently than humans.
- Penicillin was put back on the shelf for a decade because it was excreted too rapidly in a rabbit.
- Medications such as Rezulin, Propulsid, Fen-Phen, and benzbromarone tested safe in animals but killed humans.
- People who work in the animal labs have been killed by viruses such as Herpes B, which are benign in their nonhuman hosts but lethal to humans.
- Hormone Replacement Therapy (HRT) was sold to millions based on animal tests. (While HRT is appropriate for some, it increases the risk for many diseases in most.)
- The failed Alzheimer's vaccine caused some of the test patients to worsen.
- The numerous failed AIDS vaccines, some of which increased risk.
- HIV transmission through contaminated blood in France, which was caused in part because of reliance on animal models.
- Lost cures for cancers because of adverse effects or lack of efficacy in animals.
- Pharmaceutical companies have lost very large sums of money because animal tests led to development of drugs that were eventually stopped. That loss is passed on to the consumer.
- The polio vaccine was delayed because of reliance on monkeys.

(See *Animal Models in Light of Evolution* for references and further examples.)

Some people say that if we don't experiment on animals, we'll have to start experimenting on people. Is that true?

This is a scare tactic that appeals to emotion over reason, and it is frequently used by those who have a vested interest in the continuation of animal experimentation. The argument makes two erroneous assumptions: 1) animals are predictive for humans; and 2) if we stop using animals, scientists will replace them with humans in the laboratory.

No one really believes that in this day and age, scientists, cut off from the use of rats and mice, would round up humans against their will, lock them up in laboratories, and perform experiments on them. Yes, unethical human experiments have been conducted in the past (and some may be occurring today), but mostly this has little or nothing to do with the scientific issues surrounding the methodology of animal experimentation. It likely has a lot to do with old-fashioned moral weakness and evil, something well-known independently of the practice of human experimentation. It remains a fact that today ethically conducted clinical trials with proper oversight not only protect the public but also provide valuable information for researchers. It does not speak well of the policy advocates of the research community that they encourage the view that if not for rats and mice, you and your kids are next!

In another sense, every time you take a drug you are participating in an experiment. Until we have personalized medicine—where treatments will be based on each person's individual genetic makeup—anytime you take a drug, you are taking a certain risk because your exact response, compared to another person's, cannot be precisely predicted.

What breakthroughs have not relied on animals? What specific examples can you give?

Research modalities such as epidemiology, research with human tissue, advances in technology, and so forth have been responsible for many major breakthroughs.

Scanners such as the MRI and CT and the association of smoking with certain diseases are arguably the two biggest advances in medicine in the past 50 or so years. They are examples of nonanimal-based research—specifically epidemiology and technology-based advances. (Epidemiology is the study of factors affecting health and illness in populations.) The link between smoking and heart disease, and between spina bifida and folic acid deficiency, are just some of the fruits of epidemiological research. Essentially everything we know about HIV/AIDS we learned from studying humans and human tissues.

Since animal experimentation is so ingrained in the culture of science, animals will have been employed in some aspect of almost every breakthrough. Of course, it does not follow that this was the most efficient pathway. The better question is "What breakthroughs did not *need* to use animals?"

Unfortunately, the answer to that question would also require extensive inquiry into what research techniques and technologies were available at the time in question. Such inquiry would not be a very valuable addition of knowledge, as the questions science is asking today about human disease and treatment are very different than the ones asked even 50 years ago.

A study [68] in which physicians were given a list of 30 medical advances and asked to rank them in order of importance underscores the power of technology in improving patient care. Of the nine examples of technological innovations listed among the 30 medical advances, the physicians ranked eight of them in the top 15, and three of them in the top five (see table 4.3). Although, arguably, technology played a role in each of these advances, pure technology received a lion's share of accolades.

In that study, one of the authors, Victor R Fuchs, PhD, Professor Emeritus, Stanford University, noted that the study results might have far-reaching implications for expanding the criteria for quality assessment and shifting the allocation of research and development funds. He said the most surprising finding was "the extent to which the leading innovations were an outgrowth of the physical sciences (physics, engineering, and computer science) rather than disciplines traditionally associated with the biomedical sciences."

Aaron Fenster of the Robarts Research Institute:

> In the past decade, we have witnessed unprecedented advances in fields such as molecular biology, medical imaging, computer technology and computational techniques. Although advances in each field have provided exciting new insights and capabilities, it is at the interface between these fields that revolutionary advances are being made. In particular, the post-genomic era is providing opportunities for the convergence of these fields, enabling novel imaging technologies and techniques to play a significant role in drug discovery, functional genomics and measurement of pharmacokinetics and dynamics in target tissues. [69]

Table 4.3

Ranking of medical advances by physicians. Advances made possible mainly due to advances in technology in the area of physical as opposed to life sciences are noted with an asterisk.

MRI and CT scanners*
ACE inhibitors
Balloon angioplasty*
Statins
Mammography*
Coronary artery bypass graft*
Proton pump inhibitors and H2 blockers
Selective serotonin reuptake inhibitors (SSRIs) and new non-SSRI antidepressants
Cataract extraction and lens implant*
Hip and knee replacement*
Ultrasonography and echocardiography*
Gastrointestinal endoscopy*
Inhaled steroids for asthma
Laparoscopic surgery*
Nonsteroidal anti-inflammatory drugs and COX-2 inhibitors
Cardiac enzymes
Fluoroquinolones
New hypoglycemic agents
HIV testing and treatment
Tamoxifen
Prostate-specific antigen testing
Long-acting and local opioid anesthetics
Helicobacter pylori testing and treatment
Bone densitometry*
Third-generation cephalosporins
Calcium channel blockers
Intravenous conscious sedation
Sildenafil (Viagra)
Nonsedating antihistamines
Bone marrow transplant

Chapter 5. Animals and Specific Drugs and Diseases

Does your claim that animals are not predictive models for human drug and disease response hold up when one examines specific diseases?
Yes. In this chapter, we will examine specific cases where animals were used as predictive models for human drug and disease response. When researchers defend the value of using animals as heuristic devices, as a source for spare parts for humans, as bioreactors and so forth, we agree with them. As we have stated many times, animals can successfully be used in such endeavors. However, all too often, the legitimate uses of animals are brought forward as evidence that animal models are predictive

What was the role of animals in the development of the polio vaccine?
The story of the polio vaccine is a lengthy and complicated one; indeed, many books have been written on it [70]. Although none have analyzed the role of animals specifically, several conclusions can be drawn from the literature.

When polio was being studied during the early 1900s, the accepted method of study was using animals. Scientists used a number of different species, but finally settled on monkeys for two reasons. First, monkeys showed the same lesions in the spinal cord as humans; and secondly, they could be infected by tissue from other monkeys. Thus a constant supply of the virus was maintained.

While in retrospect some of the changes to the body were the same in humans and monkeys, there were many differences. A polio researcher named Flexner recommended, based on monkey studies, intraspinal serum injection at onset of symptoms in humans. It proved ineffective in humans. Another polio researcher named Brodie established immunity in monkeys using formalin to make an inactivated vaccine. He tested the vaccine on 3,000 children, with some contracting polio. Another researcher named Kolmer made a successful vaccine for monkeys. But when it was given to children, some of them contracted polio. Using monkeys to predict human response to polio vaccines and treatments was clearly unsuccessful.

In the final analysis, animals did provide a constant source of the virus for research, but they did not mimic humans in many important respects and were responsible for several major delays in getting the vaccine. Moreover, animals provided misleading information on the efficacy and safety of vaccines. But probably the worst thing to come out of animal studies was that scientists ignored human-based studies. Such was the climate at the time—and unfortunately it still is.

Almost 30 years after he developed the oral polio vaccine, Dr Albert Sabin discussed how misleading animal models delayed progress against the disease in testimony he gave under oath to the U.S. House of Representatives:

> Paralytic polio could be dealt with only by preventing the irreversible destruction of the large number of motor nerve cells, and the work on prevention was long delayed by the erroneous conception of the nature of the human disease based on misleading experimental models of the disease in monkeys [71].

Remarkably, Sabin later published an Op-Ed in a local North Carolina paper in which he essentially reversed his position, saying that animals were vital to the development of the polio vaccine.

Several things are significant about Sabin's reversal:

1. He retracted his original statement, which had been given under oath, in a local paper with limited circulation. One would assume that if he were to retract a statement of such magnitude, he would have selected *The New York Times* or *The Washington Post*, which are considered newspapers of record for the U.S. There would have been no doubt that these newspapers would have offered him the space to do so.
2. The validity of the retraction is doubtful, since it contradicts a previous statement under oath. Many people say one thing under oath and another when not under oath. Which does society usually believe?
3. When Sabin testified, he was alone speaking his mind. When he wrote the Op-Ed, he was aligned with some of the pro animal-model interest groups. Again we ask: which statement should society believe?

The AMA White Paper states: "Studies using animals also resulted in successful culture of the poliomyelitis virus; a Nobel Prize was awarded for this work in 1954 [72]." It is typical of the pro animal-model lobby that they attribute a Nobel Prize awarded for *in vitro* research to animal-based research. This manipulation of history by the AMA is not innocent. Unfortunately, this type of misleading propaganda costs us not only billions of dollars, but also lives, as resources are diverted from more promising areas of science.

Wasn't it through using animals that scientists developed insulin to treat diabetes?

Defenders of the animal model are fond of citing the development of insulin as support for continued animal-based studies. They assert, with justification, that insulin harvested from slaughterhouses prolonged the lives of many diabetics.

But crediting the recovery of insulin (a slaughterhouse by-product) to animal experiments is analogous to thanking the Toyota salesman for inventing the automobile. It is true that animals have figured largely in the history of diabetes research and especially insulin therapy, but their record is decidedly checkered in terms of giving results applicable to humans.

In 1869, scientists identified insulin-producing pancreatic cells that malfunction in diabetic patients. Other human pancreatic conditions, such as pancreatic cancer and pancreatitis (inflammation of the pancreas) were seen to produce diabetic symptoms, reinforcing the disease's link with the pancreas.

When animal-based researchers experimented on pancreases in animals, the animals did become diabetic. However, the animals' symptoms led to conjecture that diabetes was a liver disease, linking sugar transport to the liver and glycogen.

In the early 1920s two scientists, John Macleod and Frederick Banting, isolated and purified insulin by extracting it from a dog. For this they received a Nobel Prize. Macleod admitted that their contribution was not the discovery of insulin, but rather reproducing in the dog lab what had already been demonstrated in man. They were not obliged to extract insulin from dogs, because certainly there was ample tissue from humans. They merely did so because it was convention and thus convenient.

In that same year, Banting and Best gave dog insulin to a human patient with disastrous results. Note what Roberts said about the dog experiments in 1922: "The production of insulin originated in a wrongly conceived, wrongly conducted, and wrongly interpreted series of experiments [73]."

Using *in vitro* techniques, Banting, Best, and other scientists were able to mass-produce insulin from pig and cow pancreases collected at slaughterhouses. This is the main contribution animals have made to diabetics; their pancreases were used as a source for insulin. More recently, animal insulin has been replaced as science developed safer, human insulin.

What was the role of animals in the discovery of penicillin?

The how's and why's of the discovery and development of penicillin are hotly debated among scientists and medical historians; numerous versions have been circulated for many years. (This is true of most medical and scientific discoveries of the past.) However, there are some details of the penicillin story that seem to be factual and more or less universally agreed upon, beginning with the rediscovery of penicillin by Alexander Fleming in 1928. (It had actually been discovered in the late 1800s.)

Fleming then tested it *in vitro* and *in vivo* on rabbits and mice; he mentions the rabbits specifically in his original paper. The *in vitro* results showed promise, as did topical application on rabbits. But when given systemically, the rabbits metabolized it too rapidly and led Fleming to believe it would be useless for humans when administered systemically.

Some have criticized Fleming for not trying penicillin on humans. His reluctance was based on the rabbit study. Allen B. Weisse:

> [Fleming was discouraged about penicillin's possible use because first . . .] Third, after injection into an ear vein of a rabbit and with blood samples taken periodically thereafter for testing, it was found that penicillin was rapidly removed from the bloodstream. Samples taken at 30 minutes were found almost completely devoid of activity. Of what use might be an antibacterial agent that took several hours to act but was removed from the body within 30 minutes and inhibited by the blood with which it would obviously be mixing? [74]

Steffee of Bowman Gray School of Medicine states:

> Fleming considered penicillin a potential chemotherapeutic agent, but his early *in-vivo* investigations were discouraging. In rabbits, serum levels of penicillin dropped rapidly after parenteral administration, too fast to allow the several hours of contact with bacteria required for an effect *in vitro*. [75]

Steffee defends Fleming's setting penicillin aside based on the rabbit work, stating: "...how many therapeutic modalities with the poor *in vivo* results of Fleming's early penicillin trials would be offered continued funding today?"

Note also that Weisse defends Fleming's decision not to use more animals:

> One might well wonder why, given the uncontrolled devastation of bacterial diseases, no further experiments on animals or humans were undertaken. The rapid disappearance from the blood has already been mentioned . . . Even the choice not to use animal experiments more extensively, a routine practice of investigators on the continent, could be defended by Fleming and his group. After all, there might be differences between humans and other animals in resistance or susceptibility to different infections. [74]

Fleming continued to grow penicillin and even routinely gave it to humans as a topical treatment for infections prior to the 1940s [76-79]. Through a student of his, GG Paine, Fleming gave it to four humans suffering from ophthalmic neonatorum, an eye disease of infants. Three of them responded well [80, 81].

Human observation also encouraged British scientist Florey to continue the penicillin purification process. As Henderson wrote in the *Mayo Clinic Proceedings*:

> About that time, Florey who had been at Sheffield before his appointment at Oxford recalled Paine's (previously mentioned) successful topical treatment of ophthalmic neonatorium with a crude broth of penicillin. All these factors gave Florey and Chain hope that systematically administered penicillin might have therapeutic potential in humans. [80]

Florey and his colleague Chain conducted research with penicillin and, using basic chemistry, developed a method of extracting and purifying small amounts of penicillin. The purified product was tested on mice, resulting in cures of otherwise fatal infections.

The penicillin story is actually a good example of one of the many follies of using animals to model humans, which is: "Which animal do we believe?" Florey himself emphasized species differences when he stated:

> Mice were used in the initial toxicity tests because of their small size, but what a lucky chance it was, for in this respect man is like the mouse and not the guinea-pig. If we had used guinea-pigs exclusively we should have said that penicillin was toxic, and we probably should not have proceeded to try and overcome the difficulties of producing the substance for trial in man. [82]

Prior to Florey and Chain testing penicillin on humans, Fleming administered it systemically to a friend of his who was dying in the hospital. It was a desperate measure done out of necessity and the reason why many such advances are initially applied to humans. (World War II, for example, provided an opportune field trial for penicillin.) Weisse continues:

> In August 1942, a close personal friend of Fleming had contracted streptococcal meningitis. When conventional therapy failed and death seemed imminent, Fleming turned to Florey for help. The latter personally delivered his remaining supply of penicillin to Fleming and instructed him in the initial use of it. A dramatic cure was obtained, even the more so since penicillin was administered into the spinal canal for the first time to enhance its effectiveness. Publicity surrounding Fleming's friend led to funding to develop the drug and Fleming went down in history, rightly or wrongly, as the person responsible for penicillin [74].

Interestingly Florey, co-winner of the Nobel Prize for penicillin, administered penicillin to a sick cat at the same time Fleming was giving it to his sick friend. Florey's cat died.

Under certain circumstances, penicillin kills guinea pigs and Syrian hamsters [83, 84]. In addition, penicillin is teratogenic in rats, causing limb malformations in offspring. This is one of the problems with using animals to predict human response. If you had been Fleming, Florey, or one of the other scientists, which species would you have believed? The dead cat? The rabbit that metabolized penicillin so rapidly? The guinea pigs and hamsters it would have killed had it been tested on them? Or the mice on which it worked?

Regardless of the role played by animals in the discovery of penicillin, animals could not then, nor can they now, predict human response to drugs and disease.

What about the use of animals in the fight against cancer?

Since President Richard Nixon launched the War on Cancer in the early 1970s, tens if not hundreds of billions of dollars have been invested in cancer research. Much if not most of that money has gone to research using animal models.

Even with this tremendous expenditure, most acknowledge that the war against this most frightening disease has been a failure. Although the 5-year survival rate (the percentage of patients surviving five or more years) has increased, this is a reflection of earlier diagnosis, not of prolonged life.

An article by Leaf in *Fortune* magazine on March 22, 2004 entitled "Why We're Losing the War on Cancer" laid much of the blame for our failure in combating cancer on using animals as models for humans:

> The cancer community has published an extraordinary 150,855 experimental studies on mice, according to a search of the PubMed database. Guess how many of them have led to treatments for cancer? Very, very few. In fact, if you want to understand where the War on Cancer has gone wrong, the mouse is a pretty good place to start. Says Weinberg: "A fundamental problem which remains to be solved in the whole cancer research effort, in terms of therapies, is that the preclinical [animal] models of human cancer, in large part, stink." . . . Even more depressing is the very real possibility that reliance on this flawed model has caused researchers to pass over drugs that would work in humans. After all, if so many promising drugs that clobbered mouse cancers failed in man, the reverse is also likely: More than a few of the hundreds of thousands of compounds discarded over the past 20 years might have been truly effective agents. Roy Herbst, who divides his time between bench and bedside at M.D. Anderson and who has run big trials on Iressa and other targeted therapies for lung cancer, is sure that happens often. "It's something that bothers me a lot," he says.

Alexander Kamb of Novartis Institutes for BioMedical Research, wrote in *Nature Reviews Drug Discovery*:

> But as a rule, investigational drugs do not enter the clinic without some rationale and supporting preclinical evidence of efficacy. Given that many of these investigational anticancer drugs eventually fail, the animal models on which clinical trials are predicated must at best be limited in power, and at worst wildly inaccurate. Either the models are too simplistic or they possess hidden complexities that hamper the collection of reliable data — grist for the mill of prediction. [85]

One of the most stunning failures of the animal model approach in cancer research was its prediction that smoking was safe. Animal experiments largely failed to demonstrate a smoking cancer connection; and despite overwhelming epidemiological evidence linking lung cancer to smoking, the tobacco industry used the lack of an animal model to claim that smoking did not cause lung cancer. Clemmensen and Hjalgrim-Jensen:

> For decades the clinical observation of an association between cigarette smoking and bronchial carcinoma was subject to unfound doubt, suspicion, and outright opposition, largely because the disease had no counterpart in mice. There seemed no end of statisticians craving for more documentation, all resulting in the fateful delay of needed legislative initiatives. [86]

Northrup stated in 1957 that:

> The failure of many investigators...to induce experimental cancers [in animals], except in a handful of cases, during fifty years of trying, casts serious doubt on the validity of the cigarette-lung cancer theory. [87]

Utidjian observed in 1988:

> Surely, not even the most zealous toxicologist would deny that epidemiology, and epidemiology alone, has indicted and incriminated the cigarette as a potent carcinogenic agent, or would claim that experimental animal toxicology could ever have done the job with the same definition. [88]

As recently as 1993—the same year that smoking cost the American people 50 billion dollars in direct medical costs—William Campbell, president and CEO of Phillip Morris testified under oath:

> Q. Does cigarette smoking cause cancer?
> A. To my knowledge, it's not been proven that cigarette smoking causes cancer.
> Q. What do you base that on?
> A. I base that on the fact traditionally, there is, you know, in scientific terms, there are hurdles related to causation, and at this time there is no evidence that they have been able to reproduce cancer in animals from cigarette smoking. [89]

If we were to do weighted averages for harm done to humans as a result of relying on the animal model over human data, the smoking-cancer connection surely outweighs all other considerations in a historical analysis of using animals as predictive models. The fact that the tobacco industry continues to quote animal data when justifying their product is further evidence of how deeply ingrained is the belief by society that data from animal studies are predictive for humans.

Many studies have shown that animals are not predictive for humans either in finding chemicals that cause cancer or finding treatments. (See *Animal Models in Light of Evolution* for details.) Cancer is no different from other diseases and treatments.

How have animal models been used in research on AIDS and heart disease?

Animals have been used by scientists to study basic cellular processes in animals artificially infected with an AIDS-like virus, or genetically altered to have cancer or modified to have heart disease. But despite these interesting scientific advances, very few have had any impact on patient care.

The breakthroughs in HIV/AIDS and heart disease have come primarily from studying humans and human tissue. Technological advances have also been important. The main progress in diseases like AIDS, cancer, and heart disease has come in the form of prevention. Early detection, in part based on advances in technology and in part based on education, has also played an important role in treating disease.

Unfortunately, prevention lacks the glamour and the potential for profit offered by discovering and developing new treatments and pharmaceuticals. Were prevention taken as seriously and funded on a level with the National Institutes of Health (NIH) and similar institutions, we would see marked decreases in illness.

Doesn't the fact that most drugs used to treat cancer in dogs and cats are also used in humans invalidate your claims?

On the face of it, that would appear to be the case. But a deeper examination of the facts leads to a different conclusion. For example: How do drugs get from the idea stage to animals? How do drugs get tested for animal use? What is the market for drug use in veterinary medicine?

Until recently a vast majority of drugs for animals, including treatments for cancer, came from human medicine. Traditionally, veterinarians experimented or conducted research on animals using drugs that had already shown to be safe and effective in humans. The low level of legal liability in veterinary medicine, compared to human medicine, has always allowed veterinarians much greater latitude in experimenting with new drugs.

If a cat was going to die from cancer and a new drug had been used in humans, standard of care was such that a veterinarian could try the drug on the cat and observe what happened without a lot of legal ramifications. The results of these experiments have been widely variable, but sometimes a drug known to be effective in humans has proved effective in the same cancer or even a different disease in dogs and cats. Even then—as might be expected—dosing regimens and adverse reactions are often quite different between animals and humans.

The main point is that historically new drugs for animals were not designed for animals because the market was simply not large enough to support the research and development costs. Drugs filtered down from humans to pets. In recent years, new drugs have been designed specifically with pets in mind, but this remains the exception.

Chapter 6. Predictive Modalities, Alternatives and the Three R's

If the animal model is invalid for predicting human drug and disease response, what do you propose science use to fulfill this function?

First, let us point out that the question is phrased correctly. The question is not *What are the alternatives to using animals?* as animals are not viable for predicting human response *ipso facto* there can be no alternatives.

Second, the only way to predict drug and disease response in you is by designing and implementing tests that use your genes—not intact whole animals and not even unethical testing in other humans. You are the only predictive test for you. The bad news is that not all the genes involved in drug and disease response have been found; therefore, as of this writing medical science only has a few tests that can be performed using your genes. The good news is that these tests work very well and save lives every day. However, many more such tests need to be invented before we truly have gene-based medicine.

When you hear that animal testing is done to make the drug supply safer, what the honest person means by that is that animal tests are done in the hopes of making Phase I clinical trials safer for humans. Phase I is when human volunteers are given a drug in large part simply to test for safety. No honest person claims that animal testing is performed in order to protect the average consumer, even though one certainly hears this claim from lobbyists. So you need to make sure you are not tricked into making a false comparison, asking drug companies to invent predictive tests for humans before they give up animal testing designed to protect consumers. That is not what the tests are designed for, even if they were viable—which of course they are not.

So what are the alternatives to using animals in general?

There are many valid research modalities that are providing scientists the answers they need in the search for treatments and cures for human disease. Here are just a few:

In vitro (test tube) research on living tissue has been instrumental for many of the great discoveries that have advanced medicine. Though human tissue has not always been employed, it could have been because it has always been in ample supply. Blood, tissues, and organ cultures are ideal test-beds.

Epidemiology is the study of populations of humans to determine factors that could account for the prevalence of the disease within a population or for their disease immunity. Combined with genetic research and other nonanimal methods enumerated here, it provides very accurate information about whole systems.

Bacteria, viruses, and fungi reveal basic cellular and genetic properties.

Autopsy and cadavers are used for clarifying disease and teaching operating techniques such as fracture fixation, spine stabilization, ligament reconstruction, and other procedures. Physical models can be made for studying the wear on joints and other physiological matters of interest.

Genetic research has elucidated many genes that are responsible for specific diseases. Physicians can now ascertain their patients' predisposition to certain diseases, which allows them to monitor individuals with greater focus and suggest optimal nutrition, lifestyle changes, and medications.

Clinical research on patients shows how humans respond to different treatments and determines whether or not one treatment is superior to another. We can attribute our fundamental knowledge of disease and hospital care to clinical research.

Post-marketing drug surveillance (PMDS) is the reporting process whereby every effect and side effect of a new medication is reported to a monitoring agency, such as the FDA. Despite its obvious benefits, PMDS is practiced rather erratically at the present time, as reporting methods are neither easy to implement nor enforced by the government. It is an underutilized opportunity that should be further explored.

Mathematical and computer modeling is a complex research method that employs mathematics to simulate living systems and chemical reactions.

Technology is largely responsible for the high standard of care we receive today. MRI scanners, CT scanners, PET scanners, X-rays, ultrasound, blood gas analysis machines, blood chemistry analysis machines, pulmonary artery catheters, arterial catheters, microscopes, monitoring devices, lasers, anesthesia machines and monitors, operating room equipment, computer based equipment, sutures, the heart-lung machine, pacemakers, electrocardiograms, electroencephalograms, bone and joint replacements, surgical staplers, laparoscopic surgery, the artificial kidney machine, and many more are examples of technological breakthroughs.

Today we also have stem cell research, gene-based medical research such as pharmacogenetics, toxicogenomics, systems biology, and other areas to study. Another important but oft-overlooked area of study is evolutionary biology. More emphasis needs to be placed on the study of evolution, the place of evolution in disease, and the implications of evolution for disease research and treatment.

When discussing animals as surrogates for humans in drug testing and disease research, society needs ways to test and conduct research that have a high predictive value for humans. We refer to these research methods and tests as *predictive modalities.* To call these predictive modalities *alternatives* is to misuse the word.

Why aren't these predictive modalities considered alternatives to the animal model?

Because an alternative implies that the original modality is a viable one, and that the alternative simply offers a different route to the same destination. This is not true when considering the animal model, because it is not a viable modality.

Many use the word *alternative* to mean any test that does not harm animals. The problem with this is that it validates using animals as predictive models for humans. Looking for alternatives to tests that don't work in the first place has been primarily a subterfuge for more animal-based studies.

So how does one properly define the word alternative?

The word *alternative* comes from the Latin *alternare*—meaning *to interchange.* According to *The New Oxford American Dictionary* it means: "One of two or more available possibilities." The important thing here is that it implies viability.

A scientifically invalid practice cannot be replaced with an alternative. Consider this example: Eating broccoli is *not* an alternative to eating rocks for nutrition, but it *is* an alternative to eating asparagus.

The Encarta Dictionary defines alternative as follows:

> Other possibility: something different from, and able to serve as a substitute for, something else.

Example: *You could take the bus as an alternative to driving.* Note that the original choice in this example—taking the bus—is viable.

> Possibility of choosing: the possibility of choosing between two different things or courses of action.

Example: *We gave you the alternative; you decided to stay.* Again, the original choice—leaving, presumably—is the original and viable choice.

> Option: either one of two or one of several things or courses of action to choose between.

Example: *I can't decide which of the two alternatives is worse.* Both are viable, just not great.

The Cambridge Dictionary defines alternative as follows: "Something that is different from something else, especially from what is usual, and offering the possibility of choice: an alternative to coffee." The original choice is viable—in this case, coffee.

Another example: *There must be an alternative to people sleeping on the streets.* The original is viable; in this case, people are actually sleeping on the streets.

I'm afraid I have no alternative but to ask you to leave (that is what I have to do). Again, the original is viable; in this case, staying is viable if the person behaved better.

The opposition parties have so far failed to set out an alternative strategy. The original is viable; in this case, the original strategy is viable, just not acceptable to all.

An alternative venue for the concert is being sought. The original is viable; in this case, the concert was scheduled for a certain venue and could have been held there, but now needs to be changed.

We could go to the Indian restaurant or, alternatively, we could try Italian. Again, the original choice is viable because indeed the Indian restaurant serves food.

Is there a circumstance when it is appropriate to use the word alternative when referring to nonanimal based modalities?

Yes. As we have seen earlier in this book, there are clearly scientifically viable uses for animals, such as replacing a damaged human aortic valve with a valve from a pig. Using animals as incubators or bioreactors to grow viruses is also scientifically viable. Within this context, then, it would be appropriate to say that a synthetic aortic valve is an alternative to a pig valve.

Remember, in order for something to be an alternative, the original choice or course of action must be viable. The use of animals as predictive models is not viable; therefore, a predictive modality that does not use animals is not an alternative.

If predictive modalities are not alternatives, then why do I hear so much about alternatives and the Three Rs?

The Three Rs is a concept that has been embraced by groups that would seemingly be on opposite sides of the issue: the animal experimentation community and many in the animal protection community. That fact alone should give one pause: one must question the motivation of groups that support the Three Rs when the very people whose conduct they supposedly oppose also support this concept. This does not prove malfeasance but it suggests closer examination is needed.

The Three Rs has been around for almost 50 years. In 1959, two British scientists, William Russell and Rex Burch, published the results of a systematic study they conducted on the ethical aspects of animal research and the development and progress of humane techniques in the laboratory. This study launched the concept of the Three Rs: *R*educing the number of animals used; *R*efining techniques so the animals suffer less; and *R*eplacing animal-based tests as alternatives are invented.

In the ensuing years, finding alternatives to animal tests—the *R*eplacement component of the Three Rs—has become a cottage industry consuming billions of dollars and employing thousands of people. Yet the Three Rs has been a dismal failure. More animals are used in research and testing now, and more money goes to animal-based studies, than in the 1950s and 1960s when Russell and Burch were popularizing the concept. Additionally, more animals are used now than when the Three Rs groups—the European Centre for the Validation of Alternative Methods (ECCVAM) and the Interagency Coordinating Committee on the Validation of Alternative Methods (ICCVAM)—were organized.

Why? Because the Three Rs have been applied to animal use that purports to predict human response. As we have discussed earlier, most animal use is justified by scientists to society-at-large on the grounds that it is predictive for humans. Now consider the number of people whose employment hinges on the search for alternatives to tests that don't work in the first place. It's no surprise that they are outraged whenever it is pointed out to them that if a test does not fulfill the function it was designed to fill, it should be abandoned *for that purpose* regardless of what else is or isn't available.

Waiting to abandon a test that does not work until we can find one that does (finding an "alternative") is not just a misuse of the word but utter nonsense as well. The Three Rs should never have been applied to animal use that purports to predict human response. But there are more problems with this failed concept.

What other problems do you see with the Three Rs?

To answer that, let's examine who defends the *animals as predictive models* industry. They can generally be divided into two groups. The first group is made up of the animal experimenters themselves—those who use animals in research or their representatives. They have their incomes directly linked to animal experiments. The second group is the people involved in the Three Rs industry who, like the animal experimenters themselves, have their incomes linked to strong claims about the predictive utility of animal models. Included in this latter group are those who profess to be advocates for animals but who say: "Gosh darn we just have to experiment on animals. We just have to." (This is a direct quote from the current chairman of the board of a large, prominent animal protection organization.)

What both groups have in common is the difficult problem of saying animals are capable of predicting human response while simultaneously saying that is not why they are used, since the evidence that animals cannot be used to predict human response is overwhelming.

Are you saying that in fact the very people who support the Three Rs also claim that animal models are predictive for humans?

The Humane Society of the United States (HSUS), The Fund for the Replacement of Animals in Medical Experiments (FRAME), the Johns Hopkins-based Coalition to Abolish Animal Testing (CAAT), and the UK-based Royal Society for the Prevention of Cruelty to Animals (RSPCA), all support the Three Rs and claim, either by direct statement or implication, that animal models are in fact predictive for humans.

From FRAME's website (http://www.frame.org.uk/): "However, FRAME recognises that immediate abolition of all animal experiments is not possible." Because:

> Vital medical research must continue to find treatments for diseases which lessen the quality of human and animal life. New consumer products, medicines, and industrial and agricultural chemicals must be adequately tested in order to identify potential hazards to human and animal health, and to the environment.

The implication that animal tests are predictive is obvious in the above statement.

Alan M Goldberg, PhD and director of CAAT:

> In my statement, I acknowledged that laboratory animals are—and will continue to be—necessary for much research, but I emphasized the need to minimize or eliminate pain and distress in *in vivo* experiments through the Three Rs of replacement, reduction, and refinement. [90]

David O Wiebers, MD is professor of neurology at the Mayo Clinic in Rochester, Minnesota and is the Chairman of the Board of the Humane Society of the United States (HSUS). Dr. Wiebers delivered the keynote address at the HSUS symposium held in Washington, D.C., on October 11, 1991 and stated the following:

> Let me make a few comments about alternatives to animal research. Most physicians and scientists would agree that the development of alternatives to using live animals in research is desirable. Indeed, some encouraging progress is being made in this area with regard to the use of tissue cultures and other *in vitro* testing, as well as mathematical and computer models. However, we should not be under any false illusions that all of the findings of animal research can be reproduced in a computer model or tissue culture given our current level of technology and understanding.

The official position of the HSUS is:

> We carry out our work on behalf of animals used and kept in laboratories primarily by promoting research methods that have the potential to replace or reduce animal use or refine animal use so that the animals experience less suffering or physical harm. (Replacement, reduction, and refinement are known as the Three Rs or alternative methods.) The Three Rs approach, rigorously applied, will benefit both animal welfare and biomedical progress. [15]

All of the above individuals and organizations will go to great lengths to explain to you that animal tests are predictive. [One of us (RG) was on the HSUS scientific advisory committee for years and is very familiar with their official and unofficial position: they strongly hold that animal models are predictive.]

We are not saying that all members or employees of the above mentioned organizations are disingenuous. Many people in these organizations are sincere but not familiar enough with science to understand what is really going on. We encourage the reader to explore these concepts with these organizations and come to their own conclusions.

Are there others in the animal protection movement who defend the scientific merits of using animals as predictive models?

Yes. We will look at one example in hopes of clarifying our position by comparing it with another. Before we continue, however, we alert the reader to the fact that the answer to this question requires a higher level of scientific understanding and exploration than the rest of this book. An understanding of this answer is not required for an understanding of the rest of this book, so feel free to proceed to the next question if you find this answer too daunting.

Kenneth J Shapiro, PhD of Psychologists for the Ethical Treatment of Animals states:

> People use the titular cliche, "like comparing apples and oranges", to argue that two items are so different as to be incomparable. If the apple were a model of the orange, could we learn anything about oranges from the study of apples, the basic science question; and, the applied science question, could we learn to grow them better.
>
> Upon reflection, of course, apples and oranges are alike in many ways. On a gross morphological level, both are round and of similar size, have skins, seeds and pulp; both grow on trees and have similar stages in development. On a more abstract and conceptual level, both are fruit, edible, and good for you. On a molecular level, both are chemically acidic, largely consist of water, and have similar cellular structure. They are a lot alike and a lot different.[91]

So far we completely agree. As we have said many times, animals and humans have many things in common. The level of examination is of paramount importance here and in the case of comparing apples and oranges. Shapiro continues:

> More generally, for any two things that are not identical, there exists a set of similarities and differences. That is why metaphoric or analogical talk is so rich and evocative — a person is a banana. Pairings vary in the degree of S/D [similarities/differences]. We can construct a meaningful metaphor from virtually any pairing. Many, but not all, such pairings give us insights into the meaning of the target object for which an analogy has been constructed. The mind is a computer.
>
> Philosophers use the term "incommensurable" to refer to two things so different that any comparison is misleading. Is a typical animal model and target incommensurable [with humans]?

This is a misleading question. One we encounter frequently. Are animals similar to humans or different? Are zebras black or white? Both are examples of the fallacy of false dichotomy we discuss in Chapter 8. The correct answer to the question combines both the supposed incompatible answers. Zebras are both black (in some parts) and white (in other parts), not one or the other. Animals are similar to humans and yet different. Again, it is the level of examination that is important here. Shapiro:

> Unfortunately, for those opposing animal research, they are not.

That is an unfortunate statement. A more truthful statement would be: "In some ways it is and in other ways it is not. It depends upon the level of examination and the conclusions one is attempting to draw."

Shapiro:

> For a typical animal model/modelled, similarities obtain relevant to the purpose of drawing the comparison.

In the early days of animal modeled research this was much truer than it is today. The level of examination was at the level of gross anatomy and physiology and on that level, as we have said, the similarities between species, especially mammals, outweigh the differences. But today we are studying disease and drug response at a level where even monozygotic twins are different. At this level the differences outweigh the similarities even in monozygotic twins.

The statement is still true when one is speaking of comparing the genome of an animal to that of humans (comparative genomics), or comparing disease response and treatment between species (comparative medicine). However in comparative studies one is often simply seeking information—not making predictions about species response. In terms of such predictions apples are not oranges and mice are not humans.

Shapiro:

> *But, fortunately, and more importantly, I will show that such similarities are not critical in the evaluation of the model.* In any case, how can we judge this? How much are apples and oranges similar and how much different? Are they more similar than different? Are the similarities more critical in any gains made from the model? (Emphasis added.)

It depends what you want from the model. If you want predictions, then the similarities better be causal and extensive because the small differences can be lethal. Shapiro continues:

> *I am arguing that measures of increased understanding and more effective treatment are critical to the evaluation of animal models, while measures of the degree of similarity to differences are not, and that these two sets of measures are not related in any simple way.* (Emphasis added.)

This is precisely where we disagree with Shapiro. When Shapiro states: "*. . . measures of increased understanding and more effective treatment are critical to the evaluation of animal models*" he is saying they can be used as predictive models. To be blunt, if effective treatments are forthcoming from animal models, it is because they are *similar* to humans in relevant respects, while not being sufficiently dissimilar to undermine the therapeutic analogy—that is, they lead to predictions about what will happen in humans. Everyone agrees that animal experiments can lead to increased understanding. Taxpayers and policy makers (not to mention drug companies hoping for profit and insulation from liability for injuries inflicted by dangerous and/or ineffective treatments), however, are interested in effective treatments, and it is precisely here that the issues of similarities, differences and predictive utility, comes to the fore. But let's see how Shapiro continues and see if he states this more plainly.

> In his theory of evolution, Darwin shifts from an emphasis on discontinuity to continuity. Beneath the apparent differences in the panoply of life forms, he finds a common ancestry. Although Darwin also uses differences to define species, his scientific style of thought unifies apparently disparate phenomena under general principles, to emphasise similarity and continuity. Humans are made in the image of animals other than humans, not God. The contemporary synthesis of Darwinian Theory and genetics also emphasises continuity by demonstrating *similarity at the biochemical level, in addition to the anatomical, physiological, and behavioural levels.* (Emphasis added.)

Similarity yes, but is the similarity sufficient to use animals as predictive models? This is a nice rendition of history but we fail to see how it helps us determine if animals can be useful in medical research designed to predict human response. A recurring theme in the literature on drug responses, deeply rooted in Darwinian thinking, emphasizes both within-species variation and between-species variation as factors that confound predictive utility. Darwin's point was that natural selection actually exploits and amplifies some of these very differences in the process of species-making [92].

Shapiro continues by making mention of the scientific work he believes to support his position:

> Arguably, we are on the cusp of a movement back to an emphasis on similarity through the influence of biocentrism, the ecological viewpoint, theories of systemic relations (chaos theory), cognitive ethology and genetic engineering. Together, these developments discover and literarily create animals that are more like us in *intellective and social capabilities and in physiological function.* (Emphasis added.)

A lot of science is mentioned here. What does it all mean? A focus on biocentrism merely draws our attention to biology (not to the relevance of evolved similarities and differences). Reference to the ecological point of view and chaos theory is, to say the least, embarrassing to one who, like Shapiro, wishes to downplay the importance of considerations rooted in analyses of similarity *and difference*.

In adapting to divergent ecological places in the house of nature, different species are subject to different Darwinian selection pressures, resulting in important biological—genetic—differences between them (snake venom, for example, is modified saliva). Chaos theory is the very branch of mathematics that shows how small differences (not withstanding enormous similarities) between complex systems can blossom into massive differences between them when identically stimulated. Cognitive ethology (which is not a mainstream scientific perspective) at most draws our attention to the fact that nonhuman animals may have richer cognitive lives than those permitted in traditional behavioristic estimates, and granting this much, such cognitive perspectives do not licence the inference to the conclusion that chimpanzees, for example, are humans in ape suits!

As for similarity of physiological function, such similarities exist—but they are often supported by different mechanisms (and the mechanisms are all important when metabolic effects of drugs and other substances, for example, are the issue). Consider humans, cats and pigs. They all achieve the same physiological function of being able to excrete the chemical phenol (carbolic acid). Cats do it through a conjugation (joining) reaction with sulfate, and pigs do it through a conjugation reaction using glucuronic acid. Humans achieve the same function using a combination of both methods.

Lets look at this another way. Studies of animal limbs have been made with respect to arthropods (roughly speaking, invertebrates with exoskeletons) and vertebrates. From an evolutionary perspective, modern vertebrates and invertebrates are distantly related organisms. However, as noted by Shubin, Tabin and Carroll [93], although there are some stunning similarities with respect to underlying mechanisms of limb development at the genetic level, the differences are very important (and revealing of the way evolution works). Shubin et al., propose that the genetic circuits involved in vertebrate and invertebrate limb development descend with evolutionary modification from developmental circuits in a common ancestor of both vertebrates and arthropods. Essentially, modification and re-use of features of this ancient developmental system in different contexts produces the variety of limbs seen in nature today. This is why the legs of insects are not human legs writ small! Similarities, of course, but crucial differences as well! To care about biology is to care about the details of similarity *and difference*, and not to simply sweep the latter under the rug in the name of spin and rhetorical expediency.

However, we have stated many times that animals and humans have things in common: ability to feel pain, all are composed of cells, some animals, including humans treat family members—especially offspring—differently from other members of the tribe, DNA is the method of transmitting heritable traits and so on. Our admission even includes similarity of physiological function. But this similarity does not imply predictive ability in biomedical research; the real crux of the issue of using animals in research. This is the fallacy commonly referred to as bait and switch; use the term physiology to refer to basic similarities between species, a concept with which no one can disagree. But when defending the use of animals as predictive models use the term to imply far more similarity than the previously stated use implied. Shapiro:

> Similarity between model and modelled and the closely related concept of validity are not coterminus with these critical evaluative measures. In fact, differences between the model and modelled also can provide impetus to new

> understanding and treatment innovations. The apples and oranges argument — that model and modelled are incommensurable — whether based on theoretical or empirical grounds, is not an adequate critique. Continuing argument that relies heavily on similarities versus differences is unconstructive, reducing to the proverbial half-empty/half-full bottle argument.

Shapiro is saying we can learn things about human disease from studying the differences between species—in other words, comparative medicine. This is true, and what we learn is that animals are often not like the human subjects they model. This is useful information, but hardly a help in the search for effective treatments unless it prompts detailed human studies. Indeed we have learned why humans are susceptible to HIV infection while nonhuman primates are not. This is fascinating knowledge but does not help us predict the efficacy of a new AIDS vaccine.

Shapiro:

> What are the odds that it will work — that the results also will obtain in the target? What do they need to be for this strategy of animal model to be effective? Opponents of animal research, arguing the science issue, often critically claim that the prediction rate is only 50%, a rate no better than tossing a coin.

Shapiro is correct, we maintain, based on the scientific evidence that the sensitivity of animals tests in predicting drug response is less than or equal to 50 percent. The positive and negative predictive values are actually much less.

> This is a misunderstanding of the situation. Let's say that the prediction rate is only 50% — this means that, half the time, the hypothesis developed in the model and tested in the target will not be confirmed in the target. . . .

No! That is not what we are saying. This is again an example of bait and switch. When using animals to predict drug and disease response in humans the actual positive and negative predictive values are far less than 50 percent. Thus, as a modality they are not predictive. However, when using animals to generate hypotheses, as they are used in basic research for example, even a 1 percent correlation is acceptable to many scientists, as they do not claim basic science research is predictive. If you are a scientist performing research merely to add more knowledge to the world and 1 percent of the time a hypothesis you develop from your work leads to something useful (in ways often unintended) for treating sick humans, then you can rightly claim that your research resulted in just that knowledge. You cannot claim your research in general is predictive for humans, however for all the reasons we have outlined.

Shapiro:

> This rate is not a strong critique of the animal model process unless there is a better method of generating [hypotheses]. The 50% success rate must be

> compared, not to a coin toss, but to other ways of generating hypotheses — against both non-animal models and hypotheses from clinical observation.

Untrue. Shapiro is again purposively conflating using animals as hypothesis generators with using them to predict drug and disease response. A 50% success rate as hypothesis generators is acceptable (provided what he means is that the hypothesis tests out as true for humans well) but is completely unacceptable as a positive or negative predictive value for drugs. Shapiro continues:

> If these other ways [of generating hypotheses] only produce a 25% success rate, then clearly, animal models are a more efficient way of generating hypotheses.

Shapiro is here making up statistics (which is acceptable in an example such as this is)—the claim is "fact free" as we know of no studies that have ever been performed comparing animal-generated hypothesis about human physiology, anatomy, biochemistry, disease, and drug response with other ways of generating such hypotheses. But let us think about this for a moment. Other ways of generating hypotheses would include:

1. Studying humans with autopsies.
2. Studying humans using clinical trials.
3. Studying human tissues in culture.
4. Studying human populations using epidemiology.
5. Studying humans using post-marketing drug surveillance (PMDS).
6. Clinical observation of humans.
7. Using human tissue to attempt to construct tissue platforms.
8. Studying humans using high-tech devices like PET, CT, and MRI scanners (especially when studying the brain).
9. Comparing human genomes.
10. Studying individual genomes and comparing the results with the diseases and drug responses the human actually experiences.
11. Mathematical and computer modeling based on humans.

The above are eleven ways of generating hypotheses about humans and they are all human-based. Which makes more sense, to generate hypotheses about humans using animals or using human data?

In this article, Shapiro appears to purposively conflating animals used in basic research with animals used to predict human response. Others (not Shapiro so far as we know) have even claimed that animals were never intended to be predictive of the causes and cures of human disease—to avoid the very arguments we are making here. (For example see Hicks' comments in chapter 2.) Not only is this latter claim utterly specious and false, it is also a claim that few would make in public, when the taxpayers are asked to pony up their hard earned money. The public (not to mention drug companies hoping for profit) rightly supports research into the cure for Aunt Betty's breast cancer or Uncle Bill's prostate cancer. They do not want to hear that their tax dollars (or research budgets) are being spent curing cancer in mice with little hope of yield of fruit for humans (whatever additions such research may make to the sum total of human knowledge). As we stated earlier, the rhetorical tactic of blending issues surrounding the predictive use of animals in science, with other uses of animals known to be independently valuable for other reasons, has been used for decades. The above illustrates why, as we said in Chapter 1, it is so important to clarify terms when discussing this issue. It also illustrates why, if possible, the results from scientific experiments, and scientific positions in general, should be quantified and expressed using numbers. In the case we are addressing, whether animals are predictive, the use of calculations such as sensitivity, specificity, and positive and negative predictive value should all be accessed before coming to a conclusion.

Some organizations take the very reasonable sounding position that all animal tests should be examined one at a time. What is wrong with that?

Development of a theory or far reaching hypothesis allows us to predict what will happen or explain actions. If we accept the theory of gravity, we need not look upwards whenever we drop something. Contrast that with the aforementioned groups' recommendation that all research protocols and tests using animals be examined one at a time. That is like saying all perpetual motion machines should be examined one at a time to make sure we are not missing out on free energy. If one's livelihood is dependent on campaigning against animals in research there can no better way to ensure job security.

Not to mention the fact that there is good money to be made from such endeavors. For example, Andrew Rowan of the HSUS earned a reported $188,693 in 2003. Not a bad income for charity work. Paul Irwin, Andrew Rowan's boss and David Wiebers' colleague at the HSUS in 2003 reportedly earned $324,175. Such salaries are on a par with, or above, those paid to professional lobbyists for the animal-based research industry, such as: Donna Marie Artuso of the Foundation for Biomedical Research, $137,500 and Jacqueline Calnan of Americans for Medical Progress $98,763 [94].

While earning large sums of money is not in and of itself proof of a conflict of interest, it does offer a possible motive for proceeding very slowly against the status quo.

(If the above-mentioned organizations feel we have misrepresented them, we urge them to state in plain language that they agree with us that animal testing is not predictive. We have here stated unequivocally that the Three Rs can be applied to areas where using animals is viable.)

Why has change come so slowly?

Given the fact that using animals as predictive models is such a deeply entrenched practice it is not surprising. Also one must consider that for decades poorly funded not-for-profits staffed by volunteers were taking on very powerful vested interest groups. It is no wonder change has not occurred in over a century. Things can change overnight, however, when industry decides a modality is not useful and spurs government into action.

An article by Robinson et al in the March 2008 issue of *Regulatory Toxicology and Pharmacology* concludes:

> Regulatory guidelines indicate acute toxicity studies in animals are considered necessary for pharmaceuticals intended for human use. This is the only study type where lethality is mentioned as an endpoint. The studies are carried out, usually in rodents, to support marketing of new drugs and to identify the minimum lethal dose. A European initiative including 18 companies has undertaken an evidence-based review of acute toxicity studies and assessed the value of the data generated. Preclinical and clinical information was shared on 74 compounds. The analysis indicated acute toxicity data was not used to (i) terminate drugs from development (ii) support dose selection for repeat dose studies in animals or (iii) to set doses in the first clinical trials in humans. The conclusion of the working group is that acute toxicity studies are not needed prior to first clinical trials in humans. Instead, information can be obtained from other studies, which are performed at more relevant doses for humans and are already an integral part of drug development. The conclusions have been discussed and agreed with representatives of regulatory bodies from the US, Japan and Europe.

Industry answers to stockholders and is less willing to continue techniques found cost ineffective. More importantly, they do not want to lose money they could have made if animal tests had not derailed good drugs.

Those who insist they want to eliminate the use of animals in biomedical research would be well served to take a lesson from industry, which has shown demonstrable leadership in this area. Clearly, when industry speaks, governments listen. The following quotes demonstrate industry's and scientists' realization that animal models are simply not predictive for humans:

An editorial in *Nature Medicine* February 2008:

> On 20 December, the European Union adopted the Innovative Medicines Initiative (IMI), a venture created by the European Commission and members of the European Federation of Pharmaceutical Industries and Associations (EFPIA; p. 107). The aim of this €2 billion initiative is to support the faster development of new drugs and to enhance Europe's competitiveness by giving a much-needed boost to its biopharmaceutical sector.
>
> To achieve this goal, the IMI, much like the Critical Path Initiative launched four years ago by the US Food and Drug Administration (FDA), started by asking what bottlenecks make the drug-discovery process so inefficient. Much like the Critical Path Initiative, the IMI found that the drug development industry is not particularly good at predicting the safety of molecules ready to start clinical development and is even worse at predicting their efficacy. In a 153-page long document, the IMI then outlined its Strategic Research Agenda (SRA)—a series of recommendations to address these and other roadblocks to drug development. Specifically, it made a series of recommendations across four key areas: predictability of safety, predictability of efficacy, resources for knowledge management and resources for education and training.[95]

Davis writing in *Immunity* 2008:

> How did we arrive at this state of affairs [where the clinical developments of immunology have lagged so far behind the basic science breakthroughs]? A good case can be made that the mouse has been so successful at uncovering basic immunologic mechanisms that now many immunologists rely on it to answer every question. Where it was once common to use a variety of species, there is now such an abundance of reagents available in mouse immunology that one has to have an overpowering reason to work in any other species, including humans. It also has raised the bar of evidence required for journals and grant reviews, as pointed out by Steinman and Mellman (2004) and by Hayday and Peakman (2008). This has skewed the field so much that most clinically trained immunologists keep at least a few (and usually a lot more) mice in the "back room" so that they can have a steady flow of papers, grant funding, etc., and some have abandoned human work entirely as a lost cause. But this is just the price of progress, no? Well, except that mice are lousy models for clinical studies. This is readily apparent in autoimmunity (von Herrath and Nepom, 2005) and in cancer immunotherapy (Ostrand-Rosenberg, 2004), where of dozens (if not hundreds) of protocols that work well in mice, very few have been successful in humans. Similarly, in neurological diseases, the mouse models have also been disappointing (Schnabel, 2008).
> Why has the mouse been so unsuccessful as a clinical model? . . . and third is the sheer evolutionary distance (65 million years) between mice and humans and the likelihood that the immune system of a short-lived, ground-dwelling mammal that can replicate quickly may be substantially different than that of a long-lived, somewhat higher off-the-ground mammal that replicates very slowly (and thus has more of an evolutionary investment in individual survival). In this regard, Mestas and Hughes (2004) have carefully delineated the many differences between mice and humans with respect to various immune markers, as have recent reports contrasting human versus mouse phenotypes (von Bernuth et al., 2008; Cohen et al., 2006) in specific gene deficiencies [96].

Hughes in *Nature Reviews Drug Discovery* 2008:

> Another issue is that preclinical studies do not always identify the potential for DILI. A recent study published by an industry-led initiative known as the Safety Intelligence Program (SIP) Board (http://www.biowisdom.com/files/SIP_Board_Species_Concordance.pdf) quantifies this using a collection of extracted data evidenced in both Medline abstracts and the EMEA European Public Assessment Reports showing that 38–51% of DILI in humans was not detected in preclinical tests. The SIP Board is now working with other communities to try to understand why the human hepatotoxic potential of these compounds could not be predicted. [97]

Schnabel:

> The results of drug tests in mice have never translated perfectly to tests in humans. But in recent years, and especially for neurodegenerative diseases,

mouse model results have seemed nearly useless. In the past year, for example, three major Alzheimer's drug candidates, Alzhemed (3-amino-1-propanesulphonic acid), Flurizan (tarenflurbil) and bapineuzumab, all of which had seemed powerfully effective in mouse models, have performed weakly or not at all in clinical trials involving thousands of human Alzheimer's patients.
In the case of ALS, close to a dozen different drugs have been reported to prolong lifespan in the SOD1 mouse, yet have subsequently failed to show benefit in ALS patients. In the most recent and spectacular of these failures, the antibiotic minocycline, which had seemed modestly effective in four separate ALS mouse studies since 2002, was found last year to have worsened symptoms in a clinical trial of more than 400 patients (Gordon, P. H. et al. Lancet Neurol. 6, 1045–1053 (2007)). . . .
Robert Friedlander, at Harvard Medical School's Brigham and Women's Hospital, and the lead author on the first positive study of minocycline in SOD1 mice4 defends his work, saying that three other labs independently found similar results. "The fact that ALS TDI did not reproduce these results raises questions as to their methodologies," he says. As for the failed clinical trial of minocycline, Friedlander suggests that the drug may have been given to patients at too high a dose — and a lower dose might well have been effective. "In my mind, that was a flawed study," he says.
Neurologist Jeff Rothstein, who runs a large ALS research lab at Johns Hopkins University School of Medicine in Baltimore, Maryland, says of ALS TDI, "they've done some nice statistics". But the company's failure to reproduce his lab's positive study in 2002 of the anti-inflammatory drug Celebrex (celecoxib) in SOD1 mice5 might have been due to differences in study design, he says. Rothstein says that his lab confirmed Celebrex's biological effect at reducing neuroinflammation, whereas ALS TDI didn't look for it. "Were they at variance with us because they never got biological efficacy? Hard to know," says Rothstein. Celebrex later failed in a clinical trial in ALS patients.[98]

Sharon Begley writing in *Newsweek* April 21, 2008:

The way it usually works is, the rats and mice die first. Or at least get sick first. Or at the very least, show some adverse effect first—as in, before people do. The reason countless lab animals have given their lives during the testing of experimental drugs is to allow manufacturers and regulators to see that a compound might be toxic, even deadly, before millions of people use it. And if the compound does look a little dodgy, the lab-animal tests uncover the reason—how the compound affects the liver, say, or reaches the brain. Not surprisingly, these "preclinical tests" (that is, those performed before testing on humans) were especially rigorous for botulinum. One of the deadliest poisons in nature and a possible bioterrorism agent, this neurotoxin reached the market, in very dilute doses, starting in 1989 as Botox. A big reason Botox and its cousins, such as Myobloc, were OK'd was that preclinical testing showed that after being injected, they did not travel along the body's highways—nerve cells—to the brain and spinal cord. Yes, there was some evidence the toxin slipped into the bloodstream or the lymph system, but Botox in the bloodstream cannot enter the brain, says its manufacturer.

> Oops. In a reversal of the usual sequence in science, researchers have discovered, after millions of people have received the drug, something fundamental about how Botox can act. Contrary to what turned up in preclinical testing, botulinum toxin can travel along neurons from the injection site into the brain, at least in lab animals . . . That stands in contrast to the findings of earlier studies, which suggested that the neurotoxin is completely broken down at the injection site into innocuous compounds and does not migrate beyond it—or if it does, only into the bloodstream or lymph system. [99]

Eugene C. Butcher of the Department of Pathology, Stanford University, stated in *Nature Reviews Drug Discovery* 2005:

> The focus of innovation in current drug discovery is on new targets, yet compound efficacy and safety in biological models of disease — not target selection — are the criteria that determine which drug candidates enter the clinic. We consider a biology-driven approach to drug discovery that involves screening compounds by automated response profiling in disease models based on complex human-cell systems. Drug discovery through cell systems biology could significantly reduce the time and cost of new drug development.
>
> The perceived failure of current drug discovery has generated widespread concern, and several divergent opinions about the problem and its potential solutions. Horrobin has gone so far as to liken current drug discovery to an intellectually absorbing but meaningless game, divorced from the reality of medicine. A disconnection between pharmaceutical research and successful new drug discovery is indeed apparent. Far from the explosion of new drugs predicted to follow the sequencing of the human genome, the overall rate of new drug approvals has failed to keep pace with ever-increasing spending on pharmaceutical research. Even more worrisome is the rate of approval of drugs against new targets (molecules not the targets of previous drugs): over the past decade, the entire industry has averaged only two to three small-molecule drugs against such 'innovative' targets per year. Why is there not more innovation in drug therapies? What has gone wrong? . . .
>
> Although a number of underlying problems with the current paradigm have been highlighted, here we focus on two that seem particularly crucial to the rate of innovation. First, target validation (independent of an inhibitory drug) could be fruitless. Current mouse-genetics-focused methods of target validation cannot reliably predict human biology; and even if a model is predictive of human target biology, target biology cannot reliably predict drug biology. One instance of this is the failure of antagonists of the neurokinin 1 (NK1) receptor as analgesics, but a recent review reminds us, with many examples, that mice are not men. Second, the target-specific approach is exceedingly slow: only one target can be screened at a time, which creates an intrinsic bottleneck that, especially for novel targets of questionable validity, is certainly a barrier to rapid progress. . . . [100]

Curry points out:

> The failure, in the clinic, of at least fourteen potential neuroprotective agents expected to aid in recovery from stroke, after studies in animal models had predicted that they would be successful, is examined in relation to principles of extrapolation of data from animals to humans [101].

The Committee on Toxicity Testing and Assessment of Environmental Agents, National Research Council 2006:

> All or most of the reproductive cycle is evaluated. Four types of reproductive and developmental studies are discussed here—screening-level reproductive-toxicity assays, prenatal developmental-toxicity and teratology studies, generational tests, and reproductive assessment with continuous breeding. These assays [animal and *in vitro*] are conducted because of their assumed relevance for predicting human hazard potential, but the data from such models may or may not be relevant for predicting human risk. Thus, the predictive power of the tests may be limited by differences in the underlying biology. A famous example of how species differences can be important is developmental exposure to thalidomide, to which rats are highly resistant and humans are exquisitely sensitive. [[102]p63]

Lindl et al. have tried to quantify the clinical utility of animal models:

> According to the German Animal Welfare Act, scientists in Germany must provide an ethical and scientific justification for their application to the licensing authority prior to undertaking an animal experiment. Such justifications commonly include lack of knowledge on the development of human diseases or the need for better or new therapies for humans. The present literature research is based on applications to perform animal experiments from biomedical study groups of three universities in Bavaria (Germany) between 1991 and 1993. These applications were classified as successful in the animal model in the respective publications. We investigated the frequency of citations, the course of citations, and in which type of research the primary publications were cited: subsequent animal-based studies, *in vitro* studies, review articles or clinical studies. The criterion we applied was whether the scientists succeeded in reaching the goal they postulated in their applications, i.e. to contribute to new therapies or to gain results with direct clinical impact. The outcome was unambiguous: even though 97 clinically orientated publications containing citations of the above-mentioned publications were found (8% of all citations), only 4 publications evidenced a direct correlation between the results from animal experiments and observations in humans (0.3%). However, even in these 4 cases the hypotheses that had been verified successfully in the animal experiment failed in every respect. The implications of our findings may lead to demands concerning improvement of the licensing practice in Germany [103].

From *The Scientist*, Mail, September 1, 2007:

> Lessons, limitations of animal models

Editor,
Model organisms provide essential windows into normal development. But, it is strange that despite years of failure, the National Institutes of Health continues to pour dollars into research for therapeutics using rodent models. 1 Obviously there are technical challenges involved in developing and refining human therapies, but mice appear to be a very, very poor model for human diseases. The most glaring I think is cancer research. How many times have cancer cures been observed in mice? The research money could be much better spent looking for better models that would be more appropriate for translational research.
Brian D. Ackley
University of Kansas
Lawrence, KS
bdackley@ku.edu

References
1. A Gawrylewski, "The trouble with animal models," The Scientist, 21(7):44-51, July 2007.

Editor,
We, as scientists, must admit that models are simply models. Although statistical robustness is certainly needed in animal studies, it must be accepted that models do not, and often cannot, recapitulate sophisticated human physiology. These vast differences between humans and non-primates are not identified until one examines systems at a biochemical level, and this has become a very rare event. For too long, we have studied evolution in terms of investigating "similarities" between different species. These examples gave us clues as to the existence of evolution. But the evolutionary process, by definition, actually refers to the vast differences that exist between species, and even between cells within a given species.
I suspect that even if all the animal models faithfully mimicked the actual primary defects found in human diseases that they would still fall short of mimicking the human situation. For this reason, I sometimes wonder to what extent science actually advanced (in terms of understanding human disease) during the genomics era?
Richard N. Sifers
Baylor College of Medicine
Houston, TX
rsifers@bcm.tmc.edu

Höerig and Pullman wrote in the *Journal of Translational Medicine* 2004:

While the goal of TR [translational research] is to implement in vivo measurements and leverage preclinical models that more accurately predict drug effects in humans, TR itself can be defined in many ways. At its core however, is the thesis that information gathered in animal studies can be translated into clinical relevance and vice versa, thus providing a conceptual basis for developing better drugs . . . Therefore, in taking a pragmatic or operational rather than a definitional approach, a key to a successful translation

> of non-human research to human clinical trials lies in the choice of biomarkers. While biological pathways tend to be homologous across species and more so than pharmacokinetic parameters such as absorption and clearance, animal models themselves have a poor record of predicting human disease outcome. . . . [104]

Brady writing in *Drug Discovery World* 2009:

> The effort to develop drugs that interact with human immune system (whether by accident or design) has been dogged by mismatch between the data derived from animal models (mice in particular) and that found in man. Although the mouse provides the most common models for many aspects of the human immune system, the 65 million years of divergence has introduced significant differences between these species, which can and has impeded the reliable transition of pre-clinical mouse data to the clinic. The industry is littered with examples of delays, reiterations or even abandoned drug programmes arising from poor translation of animal responses to man. This article highlights some of the species differences and forwards the rationale to utilise high resolution human immune cell assays to improve the successful transition from pre-clinical project to proof-of-concept in clinical trial. [105]

Chapter 7. Defending Prediction: A Tale of Politics, Money and Ego

Why does the use of animals as predictive models persist in drug testing and disease research?
Many factors contribute to the continued use of animals as predictive models, and little, if any have anything to do with science. The animal experimentation industry is a multi-billion dollar business, with many vested interests in both industry and academia that have much to gain by maintaining the *status quo*—and much to lose by a dramatic change in how research is conducted. Upton Sinclair in his 1935 classic *I, Candidate for Governor: And How I Got Licked* wrote: "It is difficult to get a man to understand something when his salary depends upon his not understanding it."

How would you describe the position of the scientific community relative to the value of the animal model?
Overall, there is an awakening within the scientific community that animal experiments are not accomplishing what they set out to do from the standpoint of prediction. More and more scientists are beginning to question the validity of the animal model but are reluctant to state that publicly for fear of committing career suicide. With their livelihoods and professional stature at stake, as well as those of their colleagues—not to mention the financial security of the university that employs them—most scientists stick to the "party line" that animals can predict human drug and disease response.

It is important to remember here that the biological sciences account for the vast majority of grant money for most universities and hence even someone from the chemistry or math department will hesitate to point out flaws in the money machine for the university. Until a critical mass of scientists is reached, do not expect to hear scientists, even scientists outside the biological sciences, speaking out.

How does the* publish or perish *system in academia fit into this?
At most universities in the United States, PhDs in science are promoted, and thus more highly compensated and respected, on the basis of how many papers

they publish in the scientific literature. It is a system that remains deeply entrenched in academia despite being widely criticized.

Conducting animal studies is the most efficient way to generate a large number of papers in the shortest amount of time. It is far easier and faster to crank out five papers using animals than to conduct human-based research. The five papers may contribute nothing to ease human suffering, but that has never been a requirement for promotion.

Proposals for animal experiments are also an excellent way for universities to obtain lucrative research grants, which can provide substantial revenue for their institutions. These grants generally come from the National Institutes of Health (NIH), the federal agency in charge of allocating taxpayer-generated funds for biomedical research, as well as from other government agencies and private foundations. The public—and the policy makers who appropriate taxpayer funds—are willing to fund this research because they have been led to believe that results derived from animal experiments are directly relevant for human health and well-being.

Are you implying that no one believes in the animal model—that it's all just a game?

No. There are many scientists who we would describe as true believers. They really are convinced that animal models predict human biomedical outcomes.

Sometimes it's a matter of naiveté. PhDs start out using the animal models because their professor tells them to and by the time they are ready to perform research on their own, the animal model is all they know how to use. Initially, some of these people really believed that the animal model works, in part because they do not see the results of animal studies in the clinics as physicians do. They really believe that they are helping to cure disease. Eventually, many do figure out that the animal model fails—by that time, though, they have a mortgage and three kids in college. (See comments by Dr Hicks in Chapter 2.)

That doesn't say much for those in the scientific community. Don't you think they're smarter than that?

Sometimes very well educated and very smart people say very wrong things. Clever people can make genuine mistakes. They may also voice their opinion because of ego or because they have an ulterior motive. The bottom line is that while people can make innocent mistakes, sometimes they just lie. There is a difference. Susan Jacoby writing in *The Age of American Unreason* stated:

> Junk thought should not be confused with stupidity or sheer ignorance, because it is often employed by highly intelligent people to mislead and confuse a public deficient in its grasp of logic, the scientific method, and basic arithmetic required to see through the pretensions of poorly designed studies. [[1] p229]

Sometimes, there is a difference between what the scientific evidence supports and what scientists say. The motto of the Royal Society for the Advancement of Science in London, England (the world's oldest scientific society) is *Nullius in Verba*, which translated loosely as *Don't take anyone's word for it*. In the present context, this is good advice.

History, however, is replete with examples of very smart people simply making mistakes: Martin Blaser, director of the Division of Infectious Medicine at Vanderbilt University, called Barry Marshall's claim that the bacteria *Helicobacter pylori* caused ulcers, "the most preposterous thing I have ever heard [106]." (Dr. Marshall received the Nobel Prize in Physiology or Medicine in 2005 for his role in the discovery of the connection between *H. pylori* and gastric disease, reversing decades of medical doctrine which held that stress, spicy foods, and too much acid in the stomach causes all stomach ulcers.)

Louis Agassiz, the famed paleontologist, glaciologist, and geologist who first proposed that the earth had been subject to a past ice age, denied Darwin's theory of evolution and said in 1867: "I trust to outlive this mania [[107] p13]." He wasn't the only one. The great geologist Charles Lyell also denied evolution, believing instead that there were many centers of creation where new species appeared as needed [Ibid.].

Lord Kelvin, the great mathematician and physicist who developed the Kelvin scale of absolute temperature measurement, thought the sun had not been around long enough for evolution to be the *modus operandi* of the forms we have today [Ibid p21]. Lord Kelvin (truly one of the brightest people ever to have lived) also thought that the study of physics had almost yielded essentially all truths as of 1900. Even today, there are many smart people who doubt evolution and others who underestimate the amount of truth science has yet to reveal.

Many otherwise sophisticated people in England rejected the Germ Theory of Disease, vaccines, and science as a thought process for decades in the 1800s.

Astronomer Simon Newcomb published a paper explaining why airplanes would never fly. His analysis was perfect except for the lift effect of airfoil (the reason airplanes fly)—and perhaps his timing as well. He published his analysis two months before the Wright brothers flew.

Even the brightest, most disciplined, and ambitious among us are resistant to change. If we've always done something the same way, we're unlikely to change unless forced to do so. So when scientists who've been experimenting on animals for years, and who have published the results in hundreds of articles in professional journals, are confronted with solid evidence of the futility of the animal model, it is no wonder that they either balk or dig their heels in. In that respect, they allow themselves to become victims of the system by blindly following in the footsteps of previous animal modelers.

Sometimes, too, it's a question of differentiating how animals are actually being used. If you confuse using animals as predictive models with using them as a modality for the generation of ideas, then clearly you are going to be using faulty reasoning. We often see this in what animal modelers themselves claim about animals predicting human response. It can actually be a very contentious issue because many animal modelers and their supporters claim that no one seriously believes or claims that animal models are predictive—rather, they are used merely as heuristic devices. Yet their statements in the scientific literature (some of which appear below) would seem to indicate otherwise, not to mention the fact that they use the predictive value of the animal model to justify its use to the taxpaying public and their policy makers.

For example, Gad wrote in *Animal Models in Toxicology* 2007:

> Biomedical sciences' use of animals as models to help understand and *predict* responses in humans, in toxicology and pharmacology in particular, remains both the major tool for biomedical advances and a source of significant controversy . . . by and large animals have worked exceptionally well as *predictive* models for humans-when properly used . . . Animals have been used as models for centuries to *predict* what chemicals and environmental factors would do to humans . . . This work [in 1792] consisted of dosing test animals with known quantities of agents (poisons or drugs), and included the careful recording of the resulting clinical signs and gross necropsy observations. The use of animals as *predictors* of potential ill effects has grown since that time . . . Very few are familiar enough with some of the history of toxicity testing to be able to counter with examples where it has not only accurately *predicted* a potential hazard to humans, but where research has directly benefited both people and animals. There are, however, many such examples. Demonstrating the actual benefit of toxicology testing and research with examples that directly relate to the everyday lives of most people and not esoteric, basic research findings (which are the most exciting and interesting products to most scientists) is not an easy task . . . If we correctly identify toxic agents (using animals and other *predictive* model systems) in advance of a product or agent being introduced into the marketplace or environment, generally it will not be introduced (or it will be removed) and society will not see death, rashes, renal and hepatic diseases, cancer, or birth defects, for example. [108] (Emphasis added.)

Hau, the author of a well-known handbook on animal experimentation, has observed: "A third important group of animal models is employed as predictive models. These models are used with the aim of discovering and quantifying the impact of a treatment, whether this is to cure a disease or to assess toxicity of a chemical compound[109]." Akkina is saying the same: "A major advantage with this *in vivo* system [genetically modified SCID mice] is that any data you get from SCID-hu mice is directly applicable to a human situation[110]."

Fomchenko and Holland express a similar sentiment in 2006:

> GEMs [genetically engineered mice] closely recapitulate the human disease and are used to predict human response to a therapy, treatment or radiation schedule . . . Using *in vitro* systems and *in vivo* xenograft brain tumor modeling provides a quick and efficient way of testing novel therapeutic agents and targets, knowledge from which can be translated and tested in more sophisticated GEMs that faithfully recapitulate human brain tumors and will likely result in high-quality clinical trials with satisfactory treatment outcomes and reduced drug toxicities. Additional use of GEMs to establish causal links between the presence of various genetic alterations and brain tumor initiation or

> determining their necessity for tumor maintenance and/or progression provide us with a glimpse into other important aspects of brain tumor biology. [111]

Krewski et al. of the Committee on Toxicity Testing and Assessment of Environmental Agents imply the predictability of animal models when they state:

> For the foreseeable future, some targeted testing in animals will need to continue, as it is not currently possible to sufficiently understand how chemicals are broken down in the body using tests in cells alone. These targeted tests will complement the new rapid assays and ensure the adequate evaluation of chemicals. [102]

Andrew Rowan, now of the Humane Society of the United States, stated in a book review of *Brute Science* in 1997:

> The differences in xenobiotic metabolism . . . are well known to toxicologists and are taken into consideration when trying to predict potential effects in humans. Such differences are not insuperable problems nor do they render all animal toxicology useless. [112]

This use of prediction is not confined to the scientific literature. If anything, it is even more widespread when scientists are speaking to the public. Chris Smith, a doctor and a clinical lecturer in virology at Cambridge University who hosts *The Naked Scientist* podcast, stated in the October 21, 2008 podcast:

> This week scientists have made a giant step forward really in a bit of work which might help people who are paralyzed because they have had a spinal cord injury to get moving again. They've shown this just using monkeys to start with but monkeys are very good models for how humans work so we think the same technique should work in humans[113].

While science as a whole is self-correcting and the scientific method is the best way we have of understanding the material world, scientists are still human, with all the frailties associated with being human. Many scientists have not questioned the basis of their belief in the utility of animal-based studies because it is so traditional and so deeply ingrained that it seems like a fundamental fact of life.

Isn't it true that clinicians who work with human patients often support animal-based research?

Medical doctors are human beings, like the rest of us, and like the rest of us they often cannot refrain from expressing their opinion regardless of how little they actually know about the subject.

More to the point, people who are trained to be medical doctors are taught early on to accept as fact the notion that animal models have predictive validity. By the time they start actually treating patients, they have long since stopped basing their practice of medicine on the animal model literature in favor of medical journals and studies that rely on human-based research. But since they rarely have the time or the inclination to go back and analyze what they were taught in college and the first two years of medical school, they never truly realize that the animal model they grew up with has no meaning to their daily practice of medicine. Physicians are busy healing—or at least trying to—and have little time for a critical examination of what they were taught in college.

While we sympathize, it must be said that they do have a responsibility, by virtue of the status society affords them (unlike *Joe Public,* who might get a pass due to his lack of education) to educate themselves on the topic before commenting, or should refrain from doing so.

Besides the researchers themselves, who else would you describe as having a vested interest in the continuation of research that uses animal models?

A huge and profitable business has been built around animal experimentation. Those involved in this business include animal breeders who supply the animals, as well as the manufacturers of the equipment and materials necessary to maintain them in a laboratory setting. Suppliers of cages and instruments designed specifically for use in animal experiments depend on the practice of animal experimentation for their continued growth and profitability, as do industry journals, such as *Lab Animal*, *Comparative Medicine*, *ILAR (Institute for Laboratory Animal Research) Journal,* and *Laboratory Animals Journal.*

Irwin Bross, former Director of Biostatistics at the Roswell Park Memorial Institute for Cancer Research stated: "They may claim to love truth, but when it is a matter of truth versus dollars, they love the dollars more...Money talks." And German surgeon Dr. Werner Hartinger said in 1989: "There are in fact, only two categories of doctors and scientists who are not opposed to vivisection: those who don't know enough about it, and those who make money from it."

Who else has a hand in maintaining the status quo?

Most pharmaceutical companies openly state that animal tests are not scientifically valid for predicting human response. At the same time, the Food and Drug Administration will privately confirm the inaccuracy of animal tests, but insist that Big Pharma (the large pharmaceutical companies) likes the *status quo*; hence their hands are tied. No one seems willing to state the obvious—that using the animal model to predict human response is counterproductive. Meanwhile, the pharmaceutical companies use animal testing as a way to provide evidence in court that they have done what is required to ensure a drug's safety. But using animals as predictive models remains a problem. Former U.S. Secretary of Health and Human Services Mike Leavitt stated in 2007: "Currently, nine out of ten experimental drugs fail in clinical studies because we cannot accurately predict how they will behave in people based on laboratory and animal studies [114]."

We have been saying for years that one reason it continues is legal. Alan M. Goldberg and Thomas Hartung wrote in *Scientific American* in January 2006:

> In reality, representatives of nine multinational companies revealed to Goldberg that all the firms use petri dish or nonmammalian tests, usually involving fish or worms, to decide if a chemical is safe enough to produce. Only then do they perform the life-span feeding studies [on animals] —to satisfy the company's lawyers and regulatory agencies.

There are so many reports of "promising animal experiments" in newspapers and on television broadcasts. Why would the media report them if they weren't true?

Those who have a vested interest in the practice of animal experimentation are inadvertently supported by the media through such reports. Newspapers and TV broadcasts enthusiastically report every "successful" animal experiment as a medical breakthrough, while downplaying the fact that the results may or may not be duplicated in humans. TV and newspaper reports of new drugs exaggerate their efficacy and minimize the side-effects. "Editors want the medical miracle[115]." [116] The issue here is not confined to animal experiments, for you only need to consider the breathless, uncritical reporting that surrounds so-called research into the healing power of prayer. The media loves to report that people with heart disease who go to church, for example, seem to do better than those who do not. They often fail to report the follow up studies which show that those patients who did not go to church would if they could, but were too sick to go [117].

What should I make of reports that say certain research studies prove animal experimentation is necessary?

One should remember the immortal words of Hubert H Humphrey: "The right to be heard does not include the right to be taken seriously." Whenever large sums of money are involved, you should not accept anyone's opinion at face value. Everything should be considered opinion—not fact—until proven otherwise.

Mark Twain said there are liars, damn liars, and statistics. Polls can be just another form of statistics. We have all heard polls, which show that a majority of Nobel Prize winners, or members of the National Academy of Science or some other group, endorses animal experiments or believes in animal models, or whatever. These are examples of the fallacy of appeal to authority, which we will discuss in greater detail in the next chapter on critical thought. The fact that many smart people believe something means nothing in terms of the validity of the belief. Positions must be proven to a reasonable degree.

That being said, we must point out that one of the greatest but least appreciated contributions to medicine was statistics. While there is some truth to the saying that "anything can be proved with statistics," statistics nonetheless changed the way medicine was practiced and medical research carried out. The use of statistics came of age with Sir Richard Doll of England's study of smoking and British physicians in the 1950s. The NIH followed with similar studies in the 1960s. With the advent of modern computers, statistics became an even more integral part of medical research [118]. But statistics can be difficult to interpret for many reasons.

Bertrand Russell suggested in *Sceptical Essays* three rules for determining if a proposition is true. (Actually, he said these rules might curb one's tendency for thinking intellectual rubbish.)

1. When experts are agreed, the opposite opinion cannot be held to be certain.
2. When they are not agreed, no opinion can be regarded as certain by a nonexpert.
3. When they all hold that no sufficient grounds for a positive opinion exist, the ordinary person would do well to suspend judgment.

Russell believed that if everyone lived by these rules, life as we know it would change dramatically. Russell said: "What we need is not the will to believe but the will to find out," and he noted that unfortunately the passion for a belief is usually inversely proportional to the evidence for that belief.

Is there a way I can judge the claims made by those who defend the animal model?

Michael Shermer, the founding publisher of *Skeptic* magazine (www.skeptic.com) and the author of *How We Believe* and *The Borderlands of Science,* writing in his column in *Scientific American* cautioned the nonscientific public to beware of nonsense masquerading as science [119]. He suggested that the public ask 10 questions when considering whether what someone says is true.

We agree, and present Shermer's 10 questions (in italics), along with our comments for the reader to consider about animals in science.

1. How reliable is the source of the claim?

Anyone with an advanced degree in science probably knows more about science than the average person. But humans make mistakes and sometimes lie. If possible, always check the context and facts for yourself and ask yourself if the data presented appear distorted in order to make them adhere to the author or spokesperson's agenda. Obviously, vested interest groups have historically been more likely to lie than people without a vested interest.

2. Does this source often make similar claims?

Or, to paraphrase, does the source also make outrageous claims that have no bearing in reality? Consider the following claims made by those with a vested interest in animal experimentation:

States United for Biomedical Research: "Today's scientific breakthroughs and tomorrow's treatments have come from research with animals—research that could not be accomplished in any other way. . . . [120]"

Botting and Morrison: "…we cannot think of an area of medical research that does not owe many of its most important advances to animal experiments. [121]."

Trull: "Every major medical advance of this century has depended on animal research [122]."

The Foundation for Biomedical Research:

> Virtually all medical knowledge and treatment—certainly almost every medical breakthrough of the last century—has involved research with animals. There is a

compelling reason for using animals in research. The reason is that we have no other choice...There are no alternatives to animal research." [123]

The Foundation for Biomedical Research:

> Virtually every major medical advance of the last 100 years (as well as advances in veterinary medicine) has depended on research with animals. Animal studies have provided the scientific knowledge that allows health care providers to improve the quality of life for humans and animals by preventing and treating diseases and disorders, and by easing pain and suffering. Knowledge gained from animal research has contributed immeasurably to a dramatically increased human life span. [124]

The American Medical Association:

> ...virtually every advance in medical science in the twentieth century, from antibiotics and vaccines to antidepressant drugs and organ transplantation, has been achieved either directly or indirectly through the use of animals in laboratory experiments. [72]

Sigma Xi, an international, multi-disciplinary research society: "...research with animals has made possible most of the advances in medicine that we today take for granted. . . . [125]"

Garattini and van Bekkum: "There is no question that most medical progress—perhaps all, in fact—has been attained through knowledge derived initially from experiments in various animal species. [126]"

These are not subtle claims. Whenever hyperbole and sweeping generalizations are used, one should be cautious.

3. Have the claims been verified by another source?

Who, besides those with a vested interest in animal experiments, support the claims made by animal experimenters? Shermer states, "We must ask, 'Who is checking the claims, and even who is checking the checkers?' "

4. How does the claim fit with what we know about how the world works?

The Theory of Evolution, molecular biology, and genetics predict that animals and humans will differ in subtle yet profoundly important ways. The claim that animals can predict human response is outdated, as today we know that because of minuscule differences in DNA, men often do not respond to drugs as women do, and that even monozygotic twins may respond differently to diseases and drugs.

5. Has anyone gone out of the way to disprove the claim, or has only supportive evidence been sought?

Don't be taken in by confirmatory bias (which we will cover in the next chapter on critical thought). Confirmatory bias is frequently seen in arguments.

We have never denied that animals do sometimes mirror humans. When our knowledge was infantile, the similarities may have outweighed the differences. The common practice of not publishing animal experiments that failed to yield the desired outcome actively impairs unbiased judgment.

Compare our position with the statements by the vested interest groups. They tend to parade the times animals and humans gave the same results but hide the times they gave very different results, which resulted in human harm. All evidence needs to be available and examined prior to making a decision.

6. Does the preponderance of evidence point to the claimant's conclusion or to a different one?

Supporters of animal experimentation point out times when an animal responded the same way a human reacted to a drug or disease. But they ignore the vast majority of times when the animal model failed and humans were harmed as a result. ALL the evidence must be considered, including the Theory of Evolution, complexity theory, and so forth. The preponderance of the evidence, e.g., evolution and molecular biology, supports the view that trans-species extrapolation is problematic.

7. Is the claimant employing the accepted rules of reason and tools of research, or have these been abandoned in favor of others that lead to the desired conclusion?

If the argument fails the tests of critical thought, then it is probably wrong. We will be discussing the importance of critical thought in more detail in the next chapter.

8. Is the claimant providing an explanation for the observed phenomena or merely denying the existing explanation?

It is one thing to attempt to prove an opponent wrong by merely pointing out that his explanation is incomplete. Most explanations are. But it is another thing entirely to prove your position correct.

9. If the claimant proffers a new explanation, does it account for as many phenomena as the old explanation did?

We are claiming that medical research is now to the point that animal models are not predictive. We admit that animal models helped establish the Germ Theory of Disease and other basic facts about life.

10. Do the claimant's personal beliefs and biases drive the conclusions, or vice versa?

We would paraphrase, "Do the claimants have a vested interest in the product?"

Science is based on critical thought, but underlying it all is intent. Intent lies below ground, hence can only be accessed with effort. Philosophical beliefs can also influence scientists. In the final analysis, claims about the material world must be evaluated using not only the above but also the criteria we discussed in the first two questions.

(For more on the above we recommend *Damned Lies and Statistics: Untangling Numbers from the Media, Politicians, and Activists* and *More Damned Lies and Statistics: How Numbers Confuse Public Issues,* both by Joel Best. Also, *The Nature of Scientific Evidence: Statistical, Philosophical, and Empirical Considerations* by Taper and Lele (Eds.).)

CHAPTER 8. How to Examine Retorts to What We Have Offered So Far

How should I examine the arguments you have made, as well as the rebuttals I will hear from others?

Separating truth from falsehood is difficult in these days when every expert on television knows how to *spin* because of media training and does not mind lying for their employer. No matter what the subject or controversy, there always seems to be someone with a doctorate or expertise in the area that says the opposite thing from what the first expert said. In order to separate fact from fiction, we need guidelines and rules, and fortunately there are such guidelines. Critical thought allows us to evaluate arguments and claims and increases our likelihood of finding the truth. Some of the guidelines of critical thought are challenging but it is well worth our while to take a few moments and brush up what some of us may have forgotten. Likewise, science is oft times at the bottom of many disagreements we face in society today, so we need to have at least a minimal understanding of some scientific principles.

I'm interested in learning more about the use of animals in biomedical research. So why do we have to talk so much about science, logic, and critical thought?

You will undoubtedly encounter people who disagree with us and you need to be able to evaluate their claims as well as ours. Before you can examine the specific claims made by those using animals in science, or those who deny any utility at all from the process, you need to understand the foundations: critical thought and science. Knowledge of some very basic facts about critical thought and science will help you better understand why we say what we do and why our arguments are sound. This same knowledge will help you in every aspect of your life. We believe an honorable life is inspired by love and guided by education, critical thought, and science.

What is science?

The word *science* comes from the Latin verb that means *to know*. It can be difficult to define, as it is used in various ways. Science is:

1. The pursuit of knowledge of the material world by use of the scientific method. Science is *amoral*, which means that it does not distinguish between right and wrong. Even though science is amoral, it can be used for immoral purposes, e.g., using mathematics to engage in embezzlement.
2. A body of knowledge obtained by studying the material universe. In this respect, too, science is amoral.
3. The entire enterprise of everything pertaining to, or associated with the results of the above. This includes funding agencies, as well as the projects they select to fund and study, and why. When we refer to science in this sense of the word we use the phrase *scientific enterprise.*

Essentially, science is a formalized method of understanding involving critical thought. It can be characterized by the observation, identification, description, experimental investigation, and theoretical explanation of natural phenomena (the physical universe) and it implies systematic methodology and study.

Are there spiritual, mystical, or inner ways of knowing that are superior to ordinary ways of knowing? Maybe, if one is speaking of art appreciation or one's quest for spiritual truth. But if any of these *ways of knowing* conflict with the laws of nature, such as the Second Law of Thermodynamics, they are wrong when used to describe the material world.

What is the definition of critical thought?

At its most basic, critical thought is learning *how* to think, not *what* to think. Consider the prospect of deciding one day to sit down at the piano and, without any musical training, play a Mozart symphony. Unless you were the rare individual who was born a prodigy, that would be impossible. By the same token, no one plays Mozart for years and then decides to learn the musical scale. You must first learn *how* to play; then you can choose *what* to play. Critical thinking teaches us how to evaluate arguments and claims so we can then form opinions, separate fact from fiction, and make better decisions.

Critical thought includes defining words very precisely, evaluating sentences and paragraphs to find out what the authors are and are not actually saying, and judging whether fallacious reasoning is used to prop up the argument.

In essence, critical thinking is a system that allows us to obtain the correct answer and evaluate claims and other people's opinions. It also includes accepting unwelcome revelations about one's own positions. As MN Plano noted: "Never attribute to malice what can be explained by stupidity. Don't assign to stupidity what might be due to ignorance. And try not to assume your opponent is the ignorant one—until you can show it isn't you."

I think I'm a pretty reasonable person. Am I critical thinker?

According to the National Council for Excellence in Critical Thinking, 1987:

> Critical thinking is the intellectually disciplined process of actively and skillfully conceptualizing, applying, analyzing, synthesizing, and/or evaluating information gathered from, or generated by, observation, experience, reflection,

> reasoning, or communication, as a guide to belief and action. In its exemplary form, it is based on universal intellectual values that transcend subject matter divisions: clarity, accuracy, precision, consistency, relevance, sound evidence, good reasons, depth, breadth, and fairness[127].

You are a critical thinker if you use all available tools to separate fact from fiction, such as experience, observation, reasoning, and other resources. You are a critical thinker if you insist on fairness and honesty when evaluating the positions of others. You are a critical thinker if you are skeptical but fair.

It sounds like critical thought relates not just to science but in decision making as a whole.

That's absolutely right. Using critical thought on a daily basis and making oneself familiar with the world around us is immensely important. Unless one makes a habit of thinking clearly and critically in matters of small importance, it is likely that when the time comes to consider a matter of great importance, you simply will not have the necessary skills to make the right decisions. One can far too easily believe in things that will result in suffering and disaster.

Are you saying that the lack of critical thinking has serious consequences?

Absolutely. If you, the people around you, the government, and other institutions base decisions on fallacious reasoning, flawed logic, or a misinterpretation of science, it is highly probable that life as you know it will be worse than if the principles of critical thought, logic, and science were followed. Granted, there are indeed situations where an individual behaved irrationally, such as playing the lottery by buying 1,000 tickets and selecting the same numbers on each ticket, and still got what he sought—say winning a million dollars. But such outcomes are rare. Like playing the lottery, ignoring the laws of logic and science may give you a good result occasionally, but the odds are against it.

Many of the atrocities that have occurred historically have their roots in the lack of critical thinking when the population as a whole failed to question bad premises. Combined with fallacious reasoning, such poor mental practices were used either to justify atrocities or led directly to them.

Maimonides said: "One must accept the truth from whatever source it comes." Logic, reason, science, and critical thought are the measures taken to prevent the illnesses of authoritarianism and dogmatism that often lead to persecution and tragedy. As Brand Blanshard noted: "Where great human good and ills are involved, the distortion of belief from any sort of avoidable cause is immoral, and the more immoral the greater the stakes."

Is there a way to test—and improve—my critical thinking skills?

You can use the exercises below to test your critical thinking skills right now. Try the puzzles on the next 2 pages (Figures 8.1 and 8.2.)

Which black cards must be uncovered in order to know if it is true that if there is a circle on the left of the card there is one on the right as well? (Answer appears at end of chapter.)

Figure 8.1

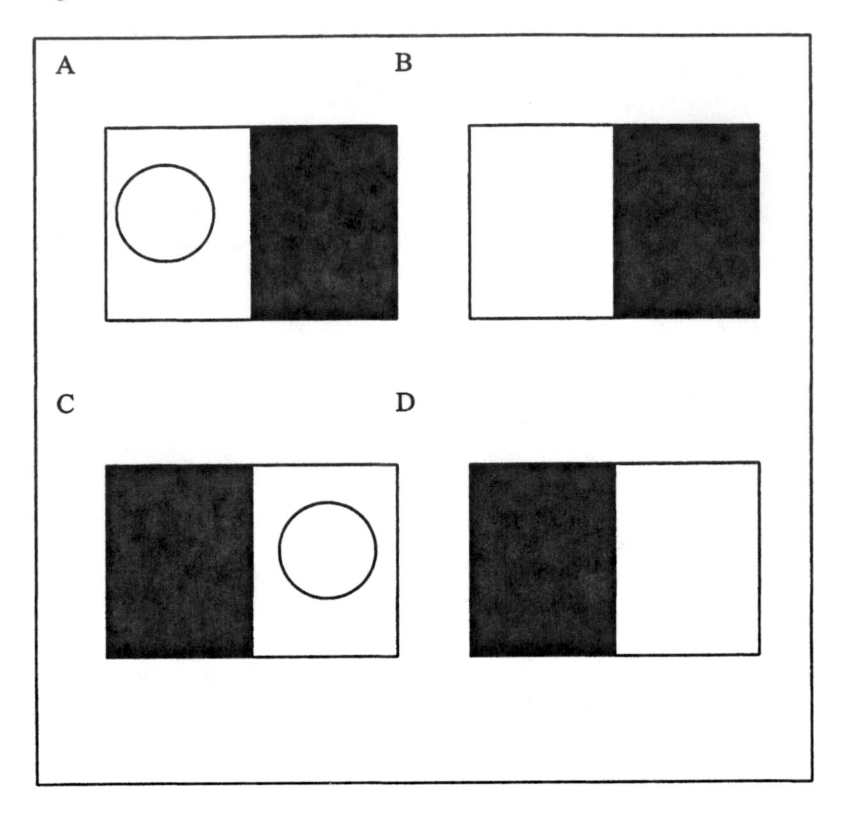

How many squares are there in the below? (Answer appears at the end of chapter.)

Figure 8.2 Squares.

If we are to examine life's options and make the proper choices, we need to be able to navigate successfully through these kinds of exercises. (More exercises like these can be found at http://www.sjsu.edu/depts/itl/graphics/adhom/fallac-x.html.)

Additional information about critical thought may be obtained on-line at http:en.wikipedia.org/wiki/Critical_thinking; Critical Thinking Web at http://philosphy.hku.hk/think/; the Critical Thinking Community at http;//www.criticalthinking.org; as well as through the following books: *How to Think About Weird Things: Critical Thinking for a New Age* by Theodore Schick and Lewis Vaughn; and *Doing Philosophy: An Introduction Through Thought Experiments* by Theodore Schick and Lewis Vaughn.

If you have no formal training in science or critical thought, you might find that you spend as much time studying these particular issues as you do on the rest of this book. If so, the investment of time will pay significant dividends, not just in your ability to understand the issue of animals in science but in your everyday life as well.

How is science related to critical thought, and why is that important to our discussion?

The scientific method combined with critical thought is the best method we have available today of thinking about the material universe. In fact, the nature of science itself and the methods it employs rest solidly on the foundations of critical thought. Take critical thought away from science and you are left with nothing more than organized voodoo.

Other than science and critical thought, there are not many viable ways to learn about our universe. Astrology, superstition, quackery, the occult, and fallacious reasoning—all examples of pseudoscience—are not science and are not reliable ways to learn about the unknown.

How does critical thought support the scientific method?

Critical thought and reason allow us to judge the probability that what we believe is true and to act accordingly. Obtaining enough evidence or data, reasoning about that data, and knowledge of science, logic, and critical thought, when all combined, can help us determine the probability that something is true or false. That is how we gain knowledge. Knowledge is power. Hence, knowledge gives us the power to accomplish our goals. As William Drummond stated: "He that will not reason is a bigot; he that cannot reason is a fool; and he that dares not reason is a slave."

Aren't logic and critical thought the same thing?

Logic is the discipline that deals with the rules of correct reasoning, and it is usually the beginning of critical thinking. Logic allows us to evaluate whether a conclusion drawn from certain premises is true. Not all things are possible. Anything that violates the laws of logic cannot be true. For example, round squares cannot exist because they are logically impossible. Nothing can be both red all over and green all over because it is logically impossible. Laws of logic are subsets of *necessary truths.*

Since logic is the science of good reasoning, anyone who argues against logic saws off the branch upon which he is sitting. One cannot use logic to refute the notion of logic. Someone who has 10 bad arguments for a given claim does not strengthen his position by virtue that he has 10 bad arguments and not just one. Ten leaky buckets hold no more water than one such bucket!

What are necessary truths?

Necessary truths are things that are absolutely true and cannot be false. Here are some examples of necessary truths:

All bachelors are not married.

2+2=4

A square has four sides.

What are the laws of logic that are subsets of these necessary truths?

There are three:

1. The law of noncontradiction. Nothing can have a property and lack it at the same time.
2. The law of identity: Everything is identical to itself.
3. The law of excluded middle: For any particular property, everything either has it or lacks it.

So, if something is logically possible, does that mean it can happen?

Not necessarily, because logic is just the beginning. Things like extra sensory perception, out-of-body experiences, and communication with the dead are not logically impossible. They may be logically possible, but they violate the laws of physics and science. Whatever violates these laws is not *logically* impossible, but *physically* impossible. Therefore, just because something is logically possible does not mean it can happen.

But what if something is logically and physically possible? Does that then mean it can happen?

Just because something is logically and physically possible does not mean it is real. Perhaps it is physically possible that we may all be energy sources for a sentient computer society, as dramatized in the movie *The Matrix*. But to live your life striving to prove that such a thing is reality is probably not the best use of your time.

This example also illustrates the important principle that just because something is physically and logically possible does not mean it is even remotely probable. Rene Descartes stated: "When it is not in our power to determine what is true, we ought to follow what is most probable." At times, this formula will lead us in the wrong direction, but most of the time it will not.

So does truth exist?

Yes, it does. As Josh Billings said: "As scarce as truth is, the supply has always been in excess of the demand." But any real or truthful situation can be represented or described differently. A map of Alaska can represent temperature or topography or roads or a myriad of other things, each real. No one map is false simply because another, formulated for a different purpose, is true. That being said, some things are just plain false.

What is an argument as it pertains to the field of logic?

Most work in the field of logic is concerned with a form of reasoning called an *argument.*

An argument consists of a set of statements called *premises.* These premises are followed by a conclusion. If the premises support the conclusion, the argument is correct. If the premises do not support the conclusion, the argument is incorrect. This is how logic allows us to evaluate whether a conclusion drawn from certain premises is true, because a valid argument will never lead from true premises to a false conclusion.

What are the different types of arguments?

In the field of logic, there are two types of arguments: *deductive* and *inductive.*

What is the deductive method?

In the deductive method, we draw conclusions by logical inference from given premises. The conclusions of deductive reasoning are referred to as *valid*, rather than *true*, because valid deductive arguments need not rest on true premises.

Can you provide an example of a logically valid deductive argument?

Premise: If A then B
Premise: A
Conclusion: B
And so:
If I am alive (A), then I am using energy (B).
I am alive (A).
I am using energy (B).

Can you provide an example of a valid argument that does not rest on true premises?

All gods live on Mount Olympus.
Michael Phelps is a god.
Therefore Michael Phelps lives on Mount Olympus.
This argument is valid because *if* the premises were true, then the conclusion would be true as well.

How do you define a deductive invalid argument?

A deductive argument is invalid when it leads from true premises to a false conclusion.

What is the inductive method?

The inductive method is a reasoning process by which one begins from a particular experience and proceeds to generalizations. Inductive arguments are not valid or truth-preserving because the truth of the premises of the inductive argument is no guarantee of the truth of the conclusion. Inductive arguments are those that may:

1. Arrive at a conclusion about a group after observing a percentage of that group.
2. Use analogical reasoning to draw conclusions about the thing being modeled. For example:

X is like Y in certain respects.
X has some property P.
It is likely Y also has property P.
3. Infer the best explanation for a hitherto unexplained phenomenon.

What good is an inductive argument if it is not a valid argument?

Remember that the word *valid*, as it is used in logic, does not mean *true*; it just means that the argument never leads from true premises to a false conclusion. So saying that an inductive argument is not *valid* is not the same thing as saying it is not true. Although an inductive argument cannot be a valid argument, it is called a strong and truth-containing argument under this condition: if the premises are true, then there is a high probability the conclusion will be true. A strong argument with true premises is called a *cogent argument.*

How are the deductive method and the inductive method used to make scientific discoveries?

Scientists use induction to come up with general theories. From these general theories, they use deductive reasoning to make new predictions. Predictions are then tested through observation and experiment. The inductive method is then used to obtain a better general theory based on those test results. Scientists would not be able to arrive at new theories using only the deductive method; on the other hand, if they only used the inductive method, they would not be able to correct and improve theories. By combining these two methods, science is able to advance.

Are deductive and inductive reasoning the only methods scientists use to make discoveries?

No. While logic and critical thought should be involved in everything we do, and in all our learning experiences, reason alone is not adequate. Without experience, reason alone is sterile. Remember, a valid argument does not have to rest on true premises, just that *if it does*, it will not lead to a false conclusion.

In addition to logic and critical thought, there are several ways scientists and others obtain information. These include experimentation, experience, appeal to authority, and intuition (also known as common sense).

Experimentation can be thought of as a test or trial done to find out something. Experiments come in two forms: 1) physical experiments; and 2) what are called thought experiments, famous examples of which were important in the development of modern physics.

Experience involves the use of our five senses—seeing, hearing, touching, tasting, and smelling (possibly aided by machines such as telescopes, X-ray machines, and genetic sequencing equipment). Experience alone, though, can be misleading. If one just stood on the surface of the earth and watched the sun, our sensory perception would lead us to assume that it rotated around the earth.

Our senses can, and do, betray us. Standing on the earth it really does appear that the sun is moving in the sky. We should rely on and believe our senses when nothing they tell us conflicts with well-known principles of science and logic, and when we have no reason to believe we are subject to an illusion. (We are referring here to an illusion such as when a stick partially immersed in water appears bent, or the big moon illusion, which occurs when the moon is close to the horizon.)

Even when our senses do tell us things that appear to conflict with science, it does not mean our senses are misleading us, but that we should seriously question what we think we are experiencing. Alan M MacRobert and Ted Schultz observed: "Our belief may predispose us to misinterpret the fact, when ideally the fact should serve as evidence upon which we base our beliefs." This does not change with greater numbers. Just because larger groups of people believe things does not give those things any more credibility.

Appeal to authority may be appropriate in some circumstances, but not many. (In fact, appeals to authority should be avoided if at all possible, as everyone has biases.) Say you have consulted three surgeons for a medical opinion. Because they do not belong to the same practice, they may be considered to be giving advice independently of each other. If each one tells you the same thing about a prospective operation, that is fairly good evidence that there is a medical consensus. However, just because all three surgeons offer the same opinion does not necessarily make it right. Twenty years ago, the medical consensus was that most stomach ulcers were due to stress. We now know that most ulcers arise from treatable bacterial infections.

Intuition or *common sense* may play a role in the formulation of explanations, given the fact that it involves varying degrees of creativity. However, explanations must be subject to testing, and it is here that appeals to intuition are very unreliable. When used appropriately, the term *common sense* refers to useful ideas and practices that have survived testing in the *laboratory of hard knocks*. Examples might include the idea that we should not leap from the tops of tall buildings, pick up very hot objects with our bare hands, or eat certain types of roots, berries, and fungus. But that is about as far as common sense will get us. Albert Einstein once said that common sense is the "collection of prejudices acquired by age eighteen".

Common sense, of itself, does not provide us the necessary intellectual resources to do calculus, modern physics, and modern physiology. It can thus be overrated. Common sense told us that man would never fly, that the earth was flat, that illness was a divine punishment, and so on. *Uncommon* sense—the ability to see things differently from other people—rightly taught us otherwise. It is this uncommon sense that contributes to the advancement of science.

You've said that good science is based on sound reasoning. How would you define bad science?

Bad science—also called junk science or pseudoscience—is characterized by vague and misleading language, exaggerations, claiming things that cannot be tested, reliance on coincidence, ignoring counter examples, unwillingness to test putative explanations, and so on. Science differs from pseudoscience in that science requires that a hypothesis or model be subject to testing in order to be proven true or false.

What is a hypothesis?

A hypothesis should do two things: 1) explain past observations, and 2) predict novel observations. For example, the Bohr model of the atom not only explained the observed stability of matter, it enabled us to make predictions about chemistry—something it was not initially formulated to do. All other things being equal, hypotheses should exhibit a variety of virtues.

What are these virtues?

These virtues are:

Conservatism. If there is a large body of knowledge for which there is very strong evidence, a new hypothesis should be logically consistent with this body of knowledge.

Falsifiability. Any scientific hypothesis should be formulated with sufficient clarity that we know, in advance of testing, what evidence would count in favor of the hypothesis, and what evidence, were it to be found, would show the hypothesis to be false.

Generalizability. A hypothesis worth its scientific salt should explain a variety of relevantly similar cases. This is important because scientific claims must be subject to independent validation by other investigators, and must be sufficiently generalizable to permit this to occur. One of the reasons that the cold fusion hypothesis was called into doubt centered on the inability of independent researchers to reproduce the alleged phenomena. The requirement of generalizability also rules out *ad hoc* explanations that apply to only one particular case, and to nothing else.

Simplicity. All other things being equal, if there are two hypotheses that explain some given phenomenon, choose the simpler one for further testing.

Modesty. Modesty is a virtue in science. It behooves us to seek mundane explanations and avoid appeals to exotic explanations where possible. Someone with the sniffles likely has a cold and not bubonic plague!

Isn't hypothesis just another word for theory?

No, and this is a very important point to understand. In everyday speech, theories are viewed as idle speculation (for example, the theory that crop circles were the result of meddling by aliens from outer space). In science, by contrast, theories are hypotheses for which there is an abundance of high quality evidence. They do not stand opposed to matters of fact; rather, theories are hypotheses that receive strong and unambiguous support from observed facts.

Scientific theories are clusters of hypotheses that have withstood rigorous testing; for example, the Theory of Evolution. Truly robust hypotheses that have withstood numerous tests, which explain past observations, and make a significant number of novel predictions that are subsequently validated, are sometimes referred to as *laws of nature.* Thus, when creationists and others opposed to good science say that evolutionary biology is *just* a theory, they are mistaking the *everyday* usage of the word with the *scientific* usage of the word.

Why is the testing of hypotheses and ideas such an important part of science?

If a hypothesis cannot be falsified, or if the adherents of the hypothesis refuse to acknowledge that, in principle it can be shown false, then the hypothesis in question is not even a candidate for inclusion in the body of scientific knowledge. No amount of special pleading for cherished hypotheses changes this fact.

In the 1700s, Joseph Priestley maintained—despite all evidence to the contrary—that fire was a special material substance known then as *phlogiston*. Priestley and other phlogiston theorists argued that this substance was released when a metal such as iron rusts, or a metal such as magnesium burns. Rusty iron was thus iron minus some of its phlogiston. The rusty piece of iron, however, weighs more than it did prior to rusting.

To save the phlogiston hypothesis, phlogiston believers argued that phlogiston had negative weight, and hence the iron gets heavier as it loses its constituent phlogiston. Using critical reasoning, the great French chemist Antoine Lavoisier argued that if phlogiston was a material substance, then like all other such substances, it had mass. Since weight is a measure of the action of gravity on mass, phlogiston must have positive, not negative weight!

It sounds like scientists do not always follow the laws of logic. Is that true?

Unfortunately, yes. As we have discussed, good science is based on sound reasoning. This is what sets good science apart from pseudoscience. However, scientists sometimes fail to follow their own rules of logic and engage in what is known as *fallacious reasoning*, which is the opposite of sound reasoning or logic. They may use fallacious reasoning simply because they don't know better. (You would be amazed at how often this happens.) Often, however, animal experimenters use fallacious reasoning to bolster their argument in those cases where the use of animals is not scientifically valid, because in these instances, science is not in their favor.

How can I recognize fallacious reasoning?

By learning the many different types of fallacies that are often employed and how to distinguish them from a valid argument. So let's take them one by one.

What is the argument from ignorance fallacy?

The argument from ignorance fallacy is used when people try to shift the burden of proof from themselves back to their opponent. For example:

Question: How do you know ESP is real?

Answer: Because no one has ever proven it isn't.

Here's the problem: Just because a claim has never been conclusively refuted does not, by virtue of that fact alone, mean that it is true. The real issue here is a question of evidence. Is there any evidence in support of ESP?

Another example that is relevant to our discussion is this argument from ignorance:

Question: How do you know animal experiments predict human outcomes?

Answer: No one has proven they do not.

Actually, we have shown that animals cannot predict human response and even if we had not, the burden of proof would be on those making the claim that they are. A final example:

Statement: I have rejected all evidence from science because science has never shown anything important to be true.

Response: Science has indeed revealed many truths that allow us to live longer and better lives.

Just because someone lacks knowledge in a specific area does not mean such knowledge does not exist.

What is the appeal to popularity fallacy?

This fallacy uses an appeal to mass sentiment, rather than reason and evidence, in order to win support for a claim or argument. Consider this statement:

> The vast majority of the American public supports animal-based research. Hence such research is justified.

This does not address whether the vast majority of the American public is correct. Recall that in earlier times, large numbers of people believed all sorts of things we now know to be false, such as the once-held belief that the earth is flat.

Here is another example from the media officials at UCLA:

> To use violent tactics aimed at halting animal research is to take away hope from millions of people with cancer, AIDS, heart disease and hundreds of other diseases[128].

The appeal to popular sentiment is obvious; everyone agrees that violence is bad and that taking away hope is bad. But there is no proof that "halting animal research" will delay the time when we see cures for AIDS, heart disease, and other afflictions.

What is the personal abuse fallacy?

The personal abuse fallacy substitutes personal attack for reasons and evidence. Say, for example, the authors state, "We think the scientific community should re-examine the role animals play in biomedical research." Opponents of our position might use the personal abuse fallacy by making such statements as, "Why do you hate babies?" Or "Why do you value rats and mice over people?"

The personal abuse fallacy is well known in courts of law, where, lacking evidence, attorneys attack witnesses on a personal level rather than addressing the issue raised by the witness' testimony.

What is the fallacy of question begging?

The fallacy of question begging involves circular argumentation in which the conclusion to be demonstrated is stated—often differently—in the premises. This term "begging the question" has a very misleading popular usage (especially in media circles) which is unrelated to the technical issue of critical thinking. People sometimes say, "But this begs the following question…" in everyday conversation. What they really mean is, "This invites the following question…" Question begging is a very specific error and is not, contrary to popular usage, an invitation for further questions.

One example of such circular argumentation is Jack asking Jill, "Can we trust the Bible when it says God exists?" Jill replies: "We may conclude that God exists because it says so in the Bible and the Bible is the inspired word of

God." In this statement, Jill assumes her conclusion. And it is very easy to prove something when you assume it from the outset.

Now consider another example of circular argumentation: An advocate of animal models states that animal models are scientifically important for human health and well-being because the Food and Drug Administration (FDA) requires them, and the FDA would not require them if they were not highly relevant for human health and well-being.

The fallacy in question here rarely appears in such an obvious form. It usually happens in the course of long and detailed argumentation, where, even in good faith, people are apt to forget exactly what they claimed. (Of course, malice is sometimes a good explanation too.)

What is the fallacy of false dichotomy?

This fallacy is also known as the black-and-white fallacy because it is based on the question, "Is a zebra black or white?" It occurs when someone offers a choice between only two claims, when these claims are not the only relevant choices. Consider this statement: "Either animal experimentation predicts human outcomes or it is totally worthless." The fact of the matter is that animal experimentation can be of great scientific value (such as in basic research) even if it does not predict human outcomes, and is only indirectly relevant to human health and well-being.

What is the composition fallacy?

The composition fallacy involves misunderstandings about the relationships between *parts* and *wholes*. Here are two examples:

A. The parts of a car are light in weight.

B. Therefore, the entire car is light in weight.

And:

A. All the players on the team are good individually.

B. Therefore, the team is good.

(This is also known as the dream team fallacy.)

What is the division fallacy?

The division fallacy is related to the composition fallacy. Essentially, it is the other side of the coin, because it involves confusion of the properties of the whole with the properties of their parts. Example:

"The government cannot support science education *and* healthcare reform *and* research into climate change, so the government cannot afford to support science education *or* healthcare reform *or* research into climate change."

What is the fallacy of equivocation?

The fallacy of equivocation occurs when a word or phrase is used differently in the premises than in the conclusion or subsequent premises. Example:

A. All stars are in outer space.

B. Lindsay Lohan is a star.

C. Therefore, Lindsay Lohan is in outer space.

In A, the word *star* means a heavenly body appearing as a bright point in the sky at night. In B, the word means an entertainer who is exceptionally well known.

Here is another example:

A. It is obligatory that we obey the laws of the land.
B. Failing to do something obligatory is a sin against God.
C. Therefore, not following the speed limit laws is a sin against God.
In this example, A uses the word *obligatory* in the legal sense of the word. In B, it is used in the moral sense of the word. Same word, but far different meanings.

Now consider this example:
A. Animal models are demonstrably useful in science education.
B. Therefore, animal models are demonstrably useful in the search for cures for cancer.

Like most words in English, the word *useful* has more than one meaning, and these different meanings must not be confused. Before we can evaluate claims for truth or falsity, we must know exactly what those claims mean. In this sense, meaning must be established prior to truth. Fallacies of equivocation are rarely as simple as the above examples might lead one to expect. As with the fallacy of question begging, fallacies of equivocation often occur in long arguments, where the confusion may well be the result of old-fashioned human fallibility.

What is the fallacy of context?

Fallacies of context are related to fallacies of equivocation. They are often committed during debates. One participant in the debate may cite the words of an authority in the field — perhaps through a quotation shown to the audience— who has made a claim that is false, or who has committed some error of reasoning. The fallacy of context is committed when, *for no other reason*, the opponent in the debate concedes that the claim was indeed made, but argues that it has been taken out of context, and that the claim, properly understood in context, makes perfect sense.

Sometimes this charge that a claim has been taken out of context is justified and can be defended—through the citation of textual evidence, for example. But when a claim is dismissed because it has been taken out of context, and no credible supporting evidence is cited, the fallacy of context has been committed.

All scientific claims are revisable in the light of new data. Scientific claims must be formulated with sufficient clarity so we know what will count as evidence in favor of the claim in question, and what will falsify the claim. When the fallacy of context is committed, it is often no more than a way to immunize a claim from falsification.

As an example, Dr X, an astrobiologist, might be cited for his claim that there is life on Mars. When it is pointed out that neither Dr X nor anyone else has provided evidence to support this claim, a defender of Dr X might argue that his claim has been taken out of context, and what he really meant was that there are beans in Boston.

Dr X's claim is now true, if indeed this was what he meant, but it is no longer an interesting biological claim worthy of funding by the National Science Foundation!

What is the irrelevant conclusion fallacy?

An irrelevant conclusion fallacy is one in which the conclusion says nothing about the premise that it is supposed to be supporting or contradicting. Instead of arguing about the issue at hand, this fallacy typically argues for something that no one disagrees with, such as babies should not be tortured. An example of the irrelevant conclusion fallacy would go something like this: Suppose a bill for housing legislation is being debated on the senate floor. A senator stands and defends the bill but says nothing about the bill's merits other than concluding that all people deserve good housing. In the current budgetary climate, we would not be surprised to find cases of state legislators arguing that university education is useless because some highly educated men and women (e.g., Wall Street bankers and CEOs) are not good members of the community.

In the following example, notice how the statements are simply irrelevant to the subject/question:

Question: Should animals be used as predictive models for humans?

Answer: Today is Tuesday, the temperature is 71 degrees, and I am 33 years old; therefore, animals can be used as predictive models for humans.

The irrelevant conclusion fallacy is painfully obvious in the above example, but slightly less so in the following:

Question: Do experiments on animals result in cures and treatments for humans?

Answer: Humans have the right to experiment on animals. Humans are superior to animals. Humans are more important than animals. Therefore animals can be used as predictive models for humans.

What is the appeal to pity fallacy?

This fallacy attempts to gain the sympathy of the listener instead of addressing the issue in question, thereby misdirecting the argument from the relevant issues to irrelevant ones.

In an appeal to pity fallacy, a lawyer argues before a jury that his client should be found not guilty of committing the twelve rape-murders that he did in fact commit because he came from a poverty-stricken, dysfunctional family and was emotionally and physically crushed by factors beyond his control.

Similarly, advocates of animal-based studies use descriptions of children with birth defects to argue that more money should be funneled to animal experiments involving the study of birth defects. Their argument ignores the question of whether or not animal experiments are efficacious in this context.

Those opposing animal use in science make an appeal to pity when they suggest we take into consideration the suffering of animals in labs and that, when such suffering is taken into account the science behind using animals cannot be correct. In fact science and ethics often come to different conclusion because they are asking different questions or addressing different aspects of an issue.

What is the fallacy of insufficient statistics?

The fallacy of insufficient statistics assumes that a few instances of success or failure are representative of all such instances. Examples:

I bought this one coat from the ABC store and it wore out too soon. Therefore I will not shop at the ABC store again because their products are faulty.

And:

Animal experiments helped prove that the heart pumps blood; therefore, all animal experiments are useful.

Cherry picking data is another example of the fallacy of insufficient statistics. If you are presented with the results of 100 cases comparing animal outcomes with humans and you only include in your analysis the 10 cases where animals and humans shared the same outcome while ignoring the 90 times where they did not, you are cherry picking the data.

What is the false cause fallacy?

This fallacy suggests that a *casual* relationship is in fact a *causal* one. Examples include:

Every time I wash my car it rains. Therefore, my washing my car causes it to rain.

Every Nobel Laureate in medicine has done research on animals; therefore, research on animals caused the individual to win the Prize.

What is the slippery slope fallacy?

This is an argument that criticizes a proposal by saying it will inevitably lead to a series of events that will eventually end in catastrophe. Example:

If we stop using dogs to predict human drug responses, the next thing you know we will be using developmentally disabled people, the elderly, and children with birth defects to predict drug response.

What is the straw man fallacy?

The straw man fallacy is an argument made whereby a person misquotes, misrepresents, exaggerates, or otherwise distorts his opponent's argument so as to make it appear ridiculous and hence easy to disprove, like knocking down a straw man. Example:

Shanks and Greek say that animals are not predictive for human drug and disease response. An advocate for animal-based research says that Shanks and Greek really think that animals are far more intelligent than humans and that animals should be in charge of humans.

Since neither Shanks nor Greek really believe what their opponent has claimed they do, their opponent has created a straw man.

What is the appeal to authority fallacy?

In an appeal to authority fallacy, a person uses claims by a person of authority to support their argument simply on the basis of the fact that the person is an expert in something. As an example, consider someone who cites a PhD physicist who says animals are predictive for humans or a PhD physiologist who says physicians use animal models to treat their patients. Just because someone with a doctorate holds that opinion in and of itself is not convincing.

Even citing a person who is an authority in a given subject does not guarantee the position is correct. Positions require proof, not opinion, whether that opinion is informed or uninformed. Linus Pauling was a brilliant chemist, but no clinical virologist; he claimed that vitamin C could prevent or alleviate the common cold—a claim which has been found to be false.

People who are antiscience also frequently use appeal to authority. Creationists will frequently quote an evolutionist who disagrees with some aspect of evolutionary theory. The creationist will say that since the person is an expert his statement proves evolution is on the same scientific grounds as creation. Or a denier of the connection between HIV and AIDS will quote a prominent scientist who, for whatever reason questions that HIV causes AIDS and use that as justification for his position. All instances such as those described commit numerous other fallacies as well, but each relies in some part on appeal to authority.

What is the appeal to fear fallacy?

This error occurs when, for example, someone says, "You had better support animal experimentation or there will be no cures when you get cancer." This statement does not address the question as to whether animal experimentation actually leads to cures for cancer. (As a matter of fact, a cluster of diseases which fall under the general heading of "cancer' has indeed been cured many times in rats and mice, but the resulting therapies have not translated into human medicine.)

What is the appeal to tradition fallacy?

This fallacy essentially boils down to the assertion that "we have always done it this way, so it must be good." Animal experimenters often cite the discoveries of many common basic physiologic properties that came from studying animals, saying that because much of our early knowledge of physiology came from studying animals, animal experimentation today remains a scientifically important practice. Animal experimentation today may indeed be scientifically important, but not simply because this is the way we have always done it. For another example, reliance on Aristotle's views on science arguably impeded the development of good effective science (in both physics and biology) for many, many hundreds of years. Abandoning Aristotelian science was in many ways the first step to real scientific wisdom.

What is a false analogy?

A false analogy is an error in which two things that may or may not be similar are portrayed as being similar. In most false analogies, there is simply not enough evidence available to support the comparison. Example:

The earth has air and water and life and Mars has air and water so it must have life.

Compare the above example with the following argument:

Mice are vertebrates and mice are mammals. Humans are vertebrates and humans are mammals. Experiments on mice show that drug X can be metabolized safely; therefore, humans can metabolize drug X safely.

The problem here is that humans and mice have taken different evolutionary trajectories since their respective lineages diverged over 70 million years ago. In the context of drug metabolism, evolutionary differences can be more important than the similarities cited. Humans and mice are not the same animal dressed up differently, differing only with respect to body mass!

What is neglect of negative instances?

This is the tendency to ignore data that would prove one wrong. The neglect of negative instances is also called *confirmatory bias*. It is human nature to remember things that correlated and forget things that did not. For example, we all remember when our horoscope said something that more or less happened. But we tend to forget all the other times when it said something would happen that did not.

Avoiding confirmatory bias is one of the most important things good science does well and pseudoscience (or junk science) does badly. For example, animal experimenters point out that penicillin cured infections in mice, but fail to mention that a decade earlier, data from rabbits suggested it would not be effective in humans.

What is an* ad hoc *hypothesis?

An *ad hoc* hypothesis—literally meaning "in this instance only"—is an example of defective thinking that involves making a hypothesis that cannot be verified independently of the phenomenon it is supposed to explain. An *ad hoc* hypothesis is used to buttress an argument that has gaping holes. For example, when someone says he knows all past perpetual motion machines have not been functional but this one is because it _____ (fill in the blank: is made from plutonium; was blessed by his wife; was made under the sign of the twins; and so forth).

In light of this discussion of science, logic, and critical thinking, what are questions that should be asked by someone thinking critically about biomedical research?

When you use the word ____ what exactly do you mean?
Do you always mean that when you use that word?
Is that meaning consistent with usual uses of the word?
Are there any exceptions to the position you just stated?
How did you come to that conclusion?
What was said, exactly, in the text?
Do the authors of the study have any interest, financial or otherwise, in the outcome of a scientific study?
Who sponsored the study and to what end?
What is the source of your information?
What is the source of information in a report of some claim?
Is the source reliable?
What assumptions is the position based on?
Suppose you are wrong?
What are the implications?
Why did you make that inference?
Where there are competing interpretations of a claim, is one more consistent with the data?
Why is this issue significant?

How could the relevant claims be tested and measured?
Are fair comparisons being made?
Is the statistic being interpreted properly?
Do the statisticians disagree about this particular study, and if so, why?
Are the results consistent with what is accepted fact and if not why not?
If animal data were used, were the data obtained before or after the human data?
Was the outcome dependent on the animals or could it have just as well come from a nonanimal source?
Does the study measure correlation or concordance or does it actually measure predictive value?
Does the study or position confuse a casual relationship with a causal one?
Has anything changed that would make the study or position invalid?
Is the claimant saying something will happen in the future or that something may happen?
Do any disinterested third parties agree or disagree with the claims made in a study?
Does a person need a certain amount of education in order to understand this particular issue?
What are the alternative explanations for the results in question?
Do the conclusions over reach what was shown in the study?

Is there anything else I should know about critical thought before I analyze arguments and claims about the use of animals in science?

Critical thought can be studied in college and thus we have just touched the surface of what it means and entails. That being said, the previous pages describe some of the things you should keep in mind when evaluating claims and arguments. We would close by reminding the reader that some people just lie because it suits them and always be careful of accepting arguments from people with a vested interest in the outcome.

Is there anything else I should know about science before I analyze arguments and claims about the use of animals in science?

First, science works.

Prior to the advancements made possible by science, people led difficult lives, and they suffered greatly. There was an abominable lack of sanitation in most cities and towns. Babies and mothers frequently died in childbirth. There were no conveniences such as electricity, refrigeration, and telephones. Farmers plowed their fields manually, and factory workers labored in dangerous and unhealthy conditions. Most disease was life threatening, and abruptly ended lives at a very young age.

Today, thanks to advancements in science, the world is a dramatically different place. We have indoor plumbing, air conditioners, refrigerators, ovens and stoves, CT, PET and MRI scanners, sterile operating rooms, anesthetics, television (for good or bad!), theaters, radio, transistors and computers, lasers, and knowledge of even life itself vis-à-vis the DNA molecule.

It's also important to remember that science is tentative, not dogmatic, which means that it only offers the best explanation for problems of interest as the current data can provide. Any scientific explanation may have to be revised in the light of new data. Then, if we are lucky, a better or more comprehensive explanation of observations may come along. New discoveries can sometimes bring about a massive reconstruction in the way we think about the world.

For more information about science in general, we suggest you visit Wikipedia at http://en.wikipedia.org/wiki/Science, and the books *Philosophy of Science: A Very Short Introduction* by Samir Okasha, and *Theory and Reality: An Introduction to the Philosophy of Science* (Science and Its Conceptual Foundations series) by Peter Godfrey-Smith.

And remember the words of Peter Medawar in *Advice to a Young Scientist*:

> I cannot give any scientist of any age better advice than this: the intensity of the conviction that a hypothesis is true has no bearing on whether it is true or not. The importance of the strength of our conviction is only to provide a proportionately strong incentive to find out if the hypothesis will stand up to critical evaluation.

Answers:

Which cards? A and D
Number of squares? 30

CHAPTER 9. Personalized Medicine

What is personalized medicine?

Personalized medicine involves the ability of physicians to treat patients based on their own unique genetic makeup. Pharmacogenomics is a part of this exciting new field of medicine. Researchers working in pharmacogenomics focus on the link between genomic variation and pharmacological properties.

How does personalized medicine differ from traditional medicine?

Historically, the practice of medicine has been based on statistics. Strange as it may seem, if you were sick, you would want to be treated as a statistic, rather than an individual, because you would be more likely to get well by being treated as a statistic. Say you were suffering from high blood pressure (HBP), and research showed that 98 percent of people with HBP responded well to medication A rather than medication B. It may turn out that you were among the very small minority of people who needed medication B to lower your HBP. More likely than not, however, you needed A. Since there was no way to determine beforehand if you were among the majority or minority, your best bet was to take medication A.

What's the downside to the traditional method?

The downside is that millions of people are adversely affected by medications. Some have an allergic reaction, while others suffer from a side effect like bone marrow suppression, liver failure, stroke, heart attack, and so forth. Science has long been searching for a way not only to match the drug to the patient, but also to identify disease risk before the disease manifests.

What has happened to make personalized medicine possible?

Personalized medicine is a direct application of the knowledge gained from the Human Genome Project, which unraveled the human genetic code.

A 13-year international effort that began in 1990 and was completed in 2003, the Human Genome Project was a huge turning point for gene-based medicine. And it was a long time coming. In 1953, James Watson and Francis Crick discovered the double helix structure of the DNA molecule, and it was Watson who later urged the U.S. government to finance the Human Genome

Project. In the end, what it accomplished has set the stage for a whole new era in modern medicine.

The successful completion of the process of mapping and sequencing the complete chemical instructions that control heredity has paved the way for scientists to focus on identifying the genetic differences that determine how individuals metabolize drugs.

Think of the entire human genome as an alphabet. Without an alphabet, it would be difficult to communicate with words. But an alphabet alone does nothing. Scientists are now finding what all the genes in the human body do, how they interact, what happens when they fail, and so forth. This is like using the alphabet to make words. Once we have a sufficient vocabulary, we can write sonnets and novels. And once scientists know enough about all our genes, they can cure diseases or even prevent them from happening in the first place.

But all this depends on first finding the structure of DNA, then decoding the genome, then mapping the genome, and so on.

How are medical scientists taking advantage of genome data?

We now know that people metabolize drugs differently—and thus have different pharmacological and toxicological responses to drugs—because of variations in their genes. A drug that is good for your mother may not be given to you because of your differences in genetic makeup. Today we even know that men differ from women in the way they respond to drugs and the diseases they suffer from. Even monozygotic twins do not always suffer from the same diseases.

By isolating the particular gene involved in metabolizing a particular drug, scientists can now predict an individual patient's ability to metabolize that drug based on a genetic profile.

How will personalized medicine improve health care delivery?

Personalized medicine will lead to genetic testing for common conditions and their treatments or cures. The cure may involve gene therapy, which will be the most efficient way to treat a disease. If you turn off the genes responsible for the disease, then it will never appear. The cure may even be applied *in utero*.

Personalized medicine will also eliminate the need for large-scale clinical trials, which take a long time to complete and drive up the costs of drug development—and ultimately the cost of drugs. Drugs will only need to be tested on individuals who have the appropriate genetic profile.

Another benefit to personalized medicine is that it has the potential to virtually eliminate the incidence of adverse reactions. Armed with precise data based on an individual's genetic profile, doctors will be able to administer the exact dosage a patient needs to gain maximum therapeutic effect. And that means reduced hospitalizations and more successful outcomes.

Most intriguing of all is the possibility of bringing back drugs that were recalled due to severe adverse effects. For example, a drug that was recalled because it causes kidney failure in 30 percent of patients could now be given safely and effectively to the other 70 percent of people not at risk for kidney failure.

Would personalized medicine reduce the use of animals in the laboratory?
Scientists working in personalized medicine go straight to the source, using human DNA to sort out how cells react to certain chemicals. Tiny computer chips using very small samples of DNA are treated with a chemical to determine how individual genes are affected.

Pharmacogenomics uses human rather than animal DNA, which eliminates animals or animal-derived tissue from the process.

When will personalized medicine become a reality?
We are already seeing it, with breast cancer being a prime example. Breast cancer treatment is now determined in part based on a patient's genetic makeup.

About 25-30 percent of breast cancer patients overexpress the *HER2* oncogene, which is a gene involved in the development of cancer. The overexpression results in an increase in the replication of the cancer cells. Physicians are now able to identify which breast cancer patients overexpress *HER2* and give them Herceptin, a monoclonal antibody that inhibits *HER2*.

There is a dual benefit here. First, Herceptin has been designed to target only the cells that need to be killed, thus eliminating the need to administer large doses of toxic drugs that kill all cells. Additionally, being able to identify those patients who will benefit from the drug enables physicians to avoid administering the drug to patients who will not benefit from it, which also saves them from being exposed to any potential side effects.

Do you see this as the future of medicine?
Absolutely.

Finding cures and treatments for human disease takes a very long time, and the result society wants are at the very end of a long chain of events. Advances in our understanding of genes and gene expression have put us at the end of the chain for some diseases and nearing the end for others. It is this gene-based medicine that pharmaceutical companies and other industries are working to develop and utilize. Rather than a one-size-fits-all approach to medicine, we will be seeing in the not-so-distant future numerous drugs for a single disease, each designed for a specific genetic makeup. This is a far cry from basing treatment decisions on another species.

If these exciting technologies were supported with research funds that are currently earmarked for *animals as predictive models*-based research, the promise of people living longer, healthier lives could be fulfilled much sooner.

Chapter 10. Final Comments

Why did Dr Greek and Dr Shanks write this book when Dr Greek has already written three other books and Dr Shanks has two other books on the same subject?

Dr Greek's previous three books—*Sacred Cows and Golden Geese* (2000), *Specious Science* (2002) (both published by Continuum International), and *What Will We Do If We Don't Experiment on Animals?* (Trafford 2004)—were all written for the general public. While all of them cover the topic of the use of animals in science, each has a slightly different focus. In addition, they are written at increasingly higher reading levels. *Sacred Cows and Golden Geese,* the Greeks' first book, is written at the simplest level of the three and covers the empirical evidence against using animals as predictive models. *Specious Science* discusses in greater detail the evidence against using animals as predictive models from the standpoint of evolutionary biology and molecular genetics. It requires a more extensive science background than *Sacred Cows and Golden Geese. What Will We Do If We Don't Experiment on Animals?* is the most challenging reading of the three and focuses on technological advancements in research modalities that do not use animals.

In addition, the Greeks made a concerted effort to cover most of the general areas of medical research in the three books. For example, pediatrics is covered in *Specious Science*, while the history of medical research is covered in *Sacred Cows and Golden Geese.* Of the three books, *Sacred Cows and Golden Geese* is geared more for the general public, so the history of medical research vis-à-vis animal use is not covered in the depth it would have been had they put it in the second book. Examining the use of animals in medical research is a long and complicated topic, and the reader should not expect to have all his questions answered in a single short book, or even a long book for that matter. But an honest, truth-seeking person will find that the three books, when taken together, challenge much that has been accepted on faith.

Dr Shanks has published two books aimed at a more scientifically and philosophically educated audience. *Brute Science* (Routledge 1996), coauthored with Hugh LaFollette, PhD, is a scientific and ethical examination of the use of animals in scientific research. Some of the key scientific arguments made in that book have been expanded and updated in Shanks' and Greek's new book *Animal Models in Light of Evolution* (Brown Walker 2009). Shanks' second book *Animals and Science: A Guide to the Debates (Controversies in Science)* (ABC Clio 2002) is similar to *Brute Science* in that it also addresses the topic to the scientifically and philosophically educated audience, dealing with the history of animal use in science, and the implications of the scientific study of animals for questions about animal cognition.

FAQs About the Use of Animals in Science is designed to be an introduction to the topic for the reader who is looking for more of a broad-brush stroke than one would find in the other five books. However, it also updates the material in the other books and thus provides a more current examination of issues of interest. Much has happened in the biological sciences that impacts this discussion since the other five books were written. Also, this book assumes much less scientific and philosophical background than the previous books. At the opposite end of the scale is *Animal Models in Light of Evolution*, which presents the topic for people with a strong background in science—not necessarily a PhD in science, but that would certainly help. Our latest book is primarily focused on scientific questions, but touches on issues of science policy too.

You say that this book is written at the simplest level, while* Animal Models in Light of Evolution *is written at the most challenging level. Can you demonstrate the difference by stating in broad outline the message of* Animal Models in Light of Evolution *at the level in which that book is written?

All vertebrates are examples of evolved complex systems. Restricting our attention to mammals, this means that humans and mice (say) have taken divergent evolutionary trajectories in which, in the course of evolutionary time (70 million years since divergence of the respective lineages) causally relevant differences have accumulated between the species to the extent that there is no reason to suppose that the latter (mice) can serve as predictive models for the former (humans). We explore this matter from the standpoint of both evidence and basic biological theory.

Living complex systems also manifest different responses to the same stimuli due to: (1) differences with respect to genes present; (2) differences with respect to mutations in the same gene (where one species has an ortholog of a gene found in another); (3) differences with respect to proteins and protein activity; (4) differences with respect to gene regulation; (5) differences in gene expression; (6) differences in protein-protein interactions; (7) differences in genetic networks; (8) differences with respect to organismal organization (humans and rats may be intact systems, but may be differently intact); (9) differences in environmental exposures; and last but not least (10) differences with respect to evolutionary histories. These are some of the important reasons why members of one species often respond differently to drugs and toxins, and experience different diseases. These are the kinds of differences that are relevant to an assessment of animals as predictive models.

Therefore, we now understand why even two very similar complex systems (e.g., a chimpanzee and a human, or even monozygotic twins) may respond differently to drugs and experience different diseases, and hence why one complex system/species cannot reliably predict responses for another.

Current biomedical research is studying disease and drug response at the level where the differences between complex systems (be they two different species or two different humans) become highly significant from a biological point of view, hence using animals (e.g. vertebrates) as predictive causal analogical models for human disease and drug testing may eventually be replaced with methods that are biomedically effective. Immense empirical evidence supports this position.

So-called less complex organisms (e.g. *C. elegans*, *E. coli*, Drosophila, and *S. cerevisiae*) are very useful, however, for discovering, among other things, genes that produce the core processes of living systems.

Animals such as vertebrates can be viably used as a modality for ideas, education, a source of spare parts, incubators, factories and growth media, for the study of diseases affecting the same species, to study basic biological principles, and axiomatically, to add knowledge to the world.

We propose: 1) that the EPA and FDA ultimately eliminate animal testing requirements. Such tests are not predictive of either safety or efficacy; 2) that NIH and other funding agencies critically evaluate claims in grant applications concerning the predictive efficacy of animal models in the contexts of environmental toxicology, and disease and drug research. Such claims in grant proposals need serious justification, and should not be pro forma add-ons to keep those with oversight responsibility happy and content; 3) measures should be taken by funding agencies to develop methods to critically assess arguments supporting the predictive utility of animal models in specific contexts; 4) the practice of dressing up basic research proposals (antecedently known to be scientifically important) as proposals likely to be of direct and immediate predictive relevance to human medicine and well-being should be critically evaluated. A good placc to start this critical process would be with the widespread use of weasel words such as *may*, *might*, and *could possibly* in research proposals.

If I read all of these books, would I gain enough knowledge to argue successfully with a vivisector?

Even with the best of intentions, you probably couldn't. But you can read and study enough to be able to present a rational, logical, reasoned, consistent, well thought out argument to open minded people. There are many books and articles available (see http://curedisease.com/resources.html), and if you study you can learn enough to make a good case for the points we make in this and other books [5, 129-131]. If you have minimal scientific background, we estimate it will take at least a year of very solid reading to learn enough about science in general and this field in particular to become acquainted with the issue. But even then don't expect to learn enough to challenge vivisectors in their own areas of expertise. Even experts cannot learn everything about every area of medical research.

People who have a financial or emotional interest in an activity and have more knowledge about that activity can steamroll opponents with less knowledge or experience. And some are not above using fallacious reasoning and worse in order to defend their position in society.

This problem is compounded when the person advocating change is not a doctor. The sheer mass of data one needs to learn in order to be competent in science usually requires eight years of formal education after high school. Physicians usually study about twelve years after high school. With this much knowledge of science, a vivisector can bulldoze his way through even outstanding arguments against him. If you can keep the vivisector to the specific issue, you may be able to win a few points. But we have never seen a layperson accomplish this. And even if you do win a few points, the vivisector will use the fallacy of appeal to authority. He will say that he is a scientist and you are not; therefore, you are wrong. And that will be the end of the argument in the minds of many listeners. This is not logical or rational, but most of the time a crowd of people will be swayed by it and you will appear to lose the argument.

That is not to say that with a background in general science and a lot of effort in learning about a scientific topic, you couldn't present extensive arguments that reasonable people will accept. Just don't expect that you'll be able to remember every medical discovery or refute every argument vivisectors make. Given enough time, such arguments can be defeated, but it is unreasonable to expect one person to be able to defeat every argument without extensive preparation.

Our advice is not to argue with experts unless you are also an expert. Let people who are themselves experts make those arguments. If challenged by a vivisector, ask him to accept an invitation to a public debate with another scientist. He will most likely not accept, and that alone will win you more points with a crowd than any arguments you could make. And, if we are wrong and he does accept the invitation, great! Let us know, and we will be happy to join the debate.

What if I already have a background in science?

Then you should be reading everything on http://curedisease.com/resources.html and *Animal Models in Light of Evolution*. Even then, debating the subject with experts will be a challenge. Think of this area of expertise as other fields where a doctorate or residency is required in order to become proficient. We have been studying this area for twenty years and are still learning new things.

Given the controversial nature of the subject, what was the response to the Greeks' first three books and Shanks' books?

In all three books, the Greeks tried to explain most of the points in language a layperson could understand. It was that approach that generated a certain level of criticism.

Scientists wanted more science, and lay people wanted less. However, all books sold well and were generally well received, with most of the vitriolic criticism coming from people with a vested interest in the status quo.

When the Greeks questioned the scientists who had expressed the most concern for how much they had "dumbed down" some of the concepts in *Sacred Cows and Golden Geese*, they were at a loss to explain themselves. The first book is deceptive in how much scientific knowledge is required to understand some of the sections. While most of it reads like a primer for the general public, the Greeks did put in sections that would appeal to those with more scientific knowledge. Many researchers, when questioned about those sections, could not provide much discourse at all. Because the material went beyond their knowledge base, their criticism proved to be disingenuous.

Brute Science by Hugh LaFollette and Niall Shanks was received favorably by philosophical readers, especially those interested in ethical and policy questions. Interest was also generated by the discussion of issues in the history of science. There was comparatively little discussion of the scientific critique of predictive modeling practices. *Brute Science*, while defending the importance of basic research using animals, was highly critical of the predictive uses of animals in drug development and toxicological research. No scientific arguments were offered refuting these claims made in the book, except the spurious criticism that no scientists really believe animals are predictively relevant for human medicine, which, if true, should lead to an immediate evaluation of claims by researchers concerning the predictive utility typically made in grant proposals. You cannot have it both ways!

Ironically, the authors have given several lectures based on the text of *Animal Models in Light of Evolution*, and most scientists in the audience found the material challenging. This is likely more of a reflection of the current, highly specialized, state of scientific knowledge, than it is of a failure to understand and communicate ideas effectively. In recent decades knowledge in the natural sciences has become so highly specialized that even within a department, say the department of biological sciences at a research university, colleagues may have trouble following the contours of each others research. These problems tend to be magnified across disciplines. Evidently the Renaissance man really does belong in the 16th and not the 21st century! What is true in the natural sciences is especially true in the context of biomedicine. Medical research has become highly specialized, and where it is multidisciplinary, this often means only that teams composed of members with highly specialized fragmentary knowledge work together.

Writing for a popular audience, as we have tried to do here, is challenging. Consider the Greeks' earlier efforts to communicate scientific ideas to a popular audience. For some scientists, the mere fact that the Greeks' books were written for the general public was infuriating enough. If anything makes some scientists mad, it is attempting to explain science in general—and especially their area of expertise—to nonscientists. Look at almost any science-oriented book written for the general public and you will find a host of scientists criticizing it. Then, in the next breath these same scientists despair that the general public does not understand science, especially the science in their field of expertise. While the authors of this book are saddened by the current state of science education (and the educational system that has failed its students' legitimate needs if they are to effectively participate as citizens in a democracy), we find it unfortunate that attempts to communicate difficult ideas to a broad audience are met with hostility from professionals. Popularizations of science cannot be a substitute for a rigorous science education, but the public has a right to know upon what its tax dollars are being spent, and whether they are getting value for money.

Doesn't some of the criticism you receive from the scientific community include the accusation that you both take the words of researchers out of context?

This is a common criticism that pervades public debate on just about any topic. It does require further exploration here. First, short of reproducing entire books and articles, we can always be accused of taking ideas out of context. This is true of anyone engaging in public debate, and not limited to discussions involving animal-based research. We can say only that we have taken reasonable steps to give authors we cite the benefit of the doubt—a courtesy rarely extended to us. If the researchers' published words come back to haunt them, perhaps they should have been more cautious when choosing their words in the first place.

Second, all written and spoken sources require interpretation. This does not mean that anything goes. But what it does mean is that it is incumbent on us to be fair and reasonable. We have done our best in this regard, while respecting the fact that there can—and should—be honest disagreements among scholars.

Third, we have to go by the statements made in the public record. If a researcher is unhappy when his words and thoughts in the public record come back to haunt him, that is hardly our fault. His unhappiness does not support the claim that he was taken out of context. If the researcher is unhappy, he should have been more cautious in the first place, especially when comments are written or made for consumption by friendly audiences. We are critics, and it is our job to root out exaggerations, false claims, and logical errors.

Fourth, and following this latter point, we have both had the experience of being told we had taken someone's words out of context because the person making the claims we criticize repudiates them or says (in private emails or phone calls) that he didn't mean his words to be taken the way we interpreted them. (Such retractions are a convenient way to reconcile oneself with one's peers.) Again, all we have to go on is the public record. We are not mind readers and thus we have no access to what authors really intended to say (but didn't, as a matter of fact).

Fifth, in the context of public debates, our opponents sometimes claim we have taken something out of context, but provide no evidence. Of itself, this accusation, unsupported by evidence, is little more than a form of personal abuse. (Don't worry, we are used to it!) (For more on the out of context criticism see "What is the fallacy of context?" in Chapter 8.)

Why do people say that you are animal rights extremists and that your organization, Americans For Medical Advancement (AFMA), is an animal rights group?

An old lawyers' adage says: "If you have them on the facts, argue the facts; if you have them on the law, argue the law; if you have them on neither, attack their character."

By using the phrase *animal rights extremists* in the same diatribe about animal experiments, pro-animal experimentation interest groups seek to avoid engaging in true debate and instead imply that those who disagree with them are anti-science, inflexible, irrational, anti-human, and dangerous. This same technique is employed when animal model advocates claim Nazis outlawed animal experimentation (and by implication, that contemporary critics of animal experimentation are fellow travelers). (See question on Nazis in Chapter 1.)

To begin with, we should remember that the intensity with which a belief is held is not proportional to the truth of the belief. Many people believe silly things, but have never analyzed their beliefs using critical thought and science. At the very least, one's position should always be based on the reasoning skills we discussed in the critical thought chapter as well as our current knowledge of the material universe. Judge our position based on those skills. (One might also bear in mind, when considering the claim that we are animal rights extremists, that one of us (NS) eats meat, is partial to eggs and veal, wears leather shoes and uses leather belts to hold his pants up, and is concerned less with animal rights than the destinations of his tax dollars.)

There are many books and websites, including those listed below that discuss this issue from the perspective of those who disagree with us. Not many discuss the scientific aspects, choosing instead to discuss other aspects of the issue, such as the rights of humans versus animals. Take some time to do some research, then judge for yourself the scientific claims that they make and that we make.

- American Physiological Association http://www.the-aps.org
- Americans for Medical Progress http://www.ampef.org
- Foundation for Biomedical Research http://www.fbrresearch.org
- NABR http://www.fbresearch.org/fbr.html
- RDS (UK) http://www.rds-online.org.uk
- The Animal Research War by P. Michael Conn and James V. Parker.

As for Americans For Medical Advancement (AFMA) (see www.curedisease.com), the organization was founded by Drs Ray and Jean Greek in 1999. The Greeks thought there was much common ground, scientifically, that all in the animal experimentation debate could and should agree upon. They created AFMA to be a mainstream, science-based research and educational group dedicated to improving policy and decision-making regarding the use of the animal model in biomedical research. The idea was that

AFMA would analyze only the scientific issues surrounding the use of animals in science.

Therefore, AFMA opposes animal models as a modality for predicting or seeking cures and treatments for human disease based on overwhelming scientific evidence that animal models are not predictive for humans while acknowledging that animals can be successfully used in science in other ways. This is still AFMA's position, as explained in this book and in *Animal Models in Light of Evolution*.

AFMA is frequently accused by those who advocate animal-based studies as being an animal rights group, and this accusation is used as an *ad hominem* attack. For example, in their book *The Animal Research War*, Conn and Parker list AFMA under animal rights organizations. The reason for this can only be that such people are trying to confuse people who are not familiar with AFMA. Our scientific arguments are sound and our position on ethics-related issues is also well known; AFMA does not oppose the use of animals in science, but does oppose the use of scientifically invalid methodologies be they animal-based or otherwise. We take no position on issues relating to the ethics of using animals.

We are also accused by the animal rights segment of society as being pro-vivisection because we acknowledge that there are uses of animals that are scientifically viable.

We are proud of our history of offending both sides in the controversy and take it as evidence that we must be doing something right.

What are some of the most important points that you hope readers will take away from this book?

Animals are used in science in a wide variety of ways. Some people object to the very notion of using animals. Others have never heard of a use to which they object. And many have doubts about some but not all uses. The philosophical spectrum is very broad when it comes to this topic, and, as we have seen in this book, so is the scientific spectrum. Animals can be used in science for the benefit of humans, but they are not predictive for drug response and disease research. If you now understand a little about why this is so, we have accomplished our purpose.

Science is the opposite of dogmatic adherence to unfounded beliefs. Whereas dogmatism demands that its constituency not question the beliefs of the system, science welcomes and even initiates questioning. (That is not to say that the particular scientists whose livelihoods you are threatening welcome it.) Followers of dogma are taught not to study it, nor examine its veracity, nor weigh whether alternative explanations better explain the system governed by the dogma. They cannot debate the fundamentals upon which the system is based. They are taught unquestioning belief, not to search for truth.

Science, on the other hand, withstands questioning from every quarter. In any forum, all experts' opinions bear consideration, and that consideration will through consensus, determine the present understanding of truth. German philosopher Jürgen Habermas stressed "the importance of public debate and rational consensus for preventing the domination of society by one group of interests. Consensus suffers inaccuracy when relevant opinions are suppressed."

The Soviet Union rejected Mendelian genetics based on fallacious reasoning, and likewise Nazi Germany rejected Einstein's theory of relativity because he was Jewish. As long as the vested interest groups control who is and who is not allowed to speak on an issue, just like Nazi Germany and the Soviet Union, only one view will be heard. Adler said in *Intellect: Mind Over Matter*: "Anyone who wishes to think rationally should have the habit of thinking coolly, with all affective feelings or sentiments and all emotions parked outside. The heat of the passions, especially if they are strong and violent bodily commotions, cannot help but cause a disturbance or even a distortion of all intellectual work."

Along the same lines, mathematician Mark Kac once said, "a proof is something that convinces a reasonable man and a rigorous proof convinces an unreasonable man." We would add one thought to that: unreasonable men may not be convinced regardless of the persuasiveness of the proof. The easiest way to make a man unreasonable is to make his livelihood dependant on a certain activity. The man whose livelihood is threatened by a new idea will not necessarily be reasonable, rational, nor able to think coolly. Not using critical thought and the best science of the day can result in odious mistakes.

As a people, it is our responsibility to understand the basics of critical thought and science and to demand that our government institute policy in accordance with that knowledge. As an individual, you've taken the first step by educating yourself about the issue. James Madison said: "A people who mean to be their own government must arm themselves with the power which knowledge gives. A popular government without popular information or means of acquiring it is but a prologue to a farce or a tragedy or perhaps both." And Thomas Jefferson stated in 1816: "If a nation expects to be ignorant and free, in a state of civilization, it expects what never was and never will be."

We hope you have learned about how animals are used in science from reading this book. Even more so, we hope you will apply the principles of critical thought in your decision-making process every day.

References

1. Jacoby S: *The Age of American Unreason.* Pantheon; 2008.
2. Ruesch H: *Slaughter of the Innocent.* Civitas Publications; 1983.
3. **Animals in Research: A Necessary Evil?** [http://www.vivisectionfraud.com/Nec%20Evil%20booklet%PDF]
4. [http://www.caringconsumer.com/resources_ingredients_list.asp]
5. Shanks N, Greek R, Greek J: **Are animal models predictive for humans?** *Philos Ethics Humanit Med* 2009, **4**:2.
6. Wall RJ, Shani M: **Are animal models as good as we think?** *Theriogenology* 2008, **69**:2-9.
7. **Memorandum of Understanding.** [http://yosemite.epa.gov/opa/admpress.nsf/bd4379a92ceceeac8525735900400c27/35995a22ceb67467852573f0006559de!OpenDocument]
8. Weatherall D: **The use of nonhuman primates in research 2006.** pp. 92. London; 2006:92.
9. Morowitz HJ: **Humans, animals, and physicians' waiting rooms.** *Hosp Pract (Off Ed)* 1989, **24**:53-54.
10. Fritzsche U: **Animal experimentation in Nazi Germany.** *Hosp Pract (Off Ed)* 1990, **25**:16, 18.
11. Kaplan AL: **Telling it like it was.** *The Hastings Center* 1990, **20**:47-48.
12. Holden C: **Universities fight animal activists.** *Science* 1989, **243**:17-19.
13. Fritzsche U: **Nazis and animal protection.** *Anthrozoos* 1992, **5**:218-219.
14. APHIS: **RATS / MICE / and BIRDS DATABASE: RESEARCHERS, BREEDERS, TRANSPORTERS, AND EXHIBITERS. A Database Prepared by the Federal Research Division, Library of Congress under an Interagency Agreement with the United States Department of Agriculture's Animal Plant Health Inspection Service.** (USDA ed. Washington, DC; 2000.
15. [http://www.hsus.org/animals_in_research/general_information_on_animal_research/frequently_asked_questions_about_animals_in_research.html]
16. **New research reveals 115 million animals used in experiments worldwide** [http://www.drhadwentrust.org/news/new-research-reveals-115-million-animals-used-in-experiments-worldwide]
17. Taylor K, Gordon N, Langley G, Higgins W: **Estimates for worldwide laboratory animal use in 2005.** *Altern Lab Anim* 2008, **36**:327-342.

18. Pennisi E: **A mouse chronology.** *Science* 2000, **288**:248-257.
19. Waltz E: **Price of mice to plummet under NIH's new scheme.** *Nat Med* 2005, **11**:1261.
20. Committee on Models for Biomedical Research Board on Basic Biology: *Commission on Life Science. National Research Council. Models for Biomedical Research: A New Perspective.* Washington, DC: National Academy Press.; 1985.
21. Nathan DG, Schechter AN: **NIH support for basic and clinical research: biomedical researcher angst in 2006.** *JAMA* 2006, **295**:2656-2658.
22. Freeman M, St Johnston D: **Wherefore DMM?** *Disease Models & Mechanisms* 2008, **1**:6-7.
23. US Congress: **Scientific Fraud and Misconduct and the Federal Response.** (Committee on Government Operations Subcommittee on Human Resources and Intergovernmental Relations ed., vol. 100th congress, April 11 edition. Washington DC: US Congress; 1988.
24. Congress U: **US Congressional Hearings on Scientific Fraud and Misconduct.** April 12, 1989.
25. Meyers MA: *Happy Accidents.* Arcade Publishing; 2007.
26. Congress U, House Committee on Science: **Restructuring the Federal Science Establishment: Hearings before the House Science Committee, 104th Cong., 1st sess. June 28, 1995.**, vol. 104th Congress. Washington DC; June 28, 1995.
27. Carroll S: *Endless Forms Most Beautiful: The New Science of Evo Devo.* WW Norton; 2006.
28. LaFollette H, Shanks N: **Two Models of Models in Biomedical Research.** *Philosophical Quarterly* 1995:141-160.
29. Kornberg A: **Science in the stationary phase.** *Science* 1995, **269**:1799.
30. Lord Rayleigh: *The Life of Sir J.J. Thomson.* Cambridge University Press; 1942.
31. **Fraud** [http:/Encarta.ms.n.com/encnet/features/dictionary/DictionaryResults.aspx?refid=1861613344]
32. Loisel S, Ohresser M, Pallardy M, Dayde D, Berthou C, Cartron G, Watier H: **Relevance, advantages and limitations of animal models used in the development of monoclonal antibodies for cancer treatment.** *Crit Rev Oncol Hematol* 2007, **62**:34-42.
33. Jochems CE, van der Valk JB, Stafleu FR, Baumans V: **The use of fetal bovine serum: ethical or scientific problem?** *Altern Lab Anim* 2002, **30**:219-227.
34. PCRM: *Good Medicine* 2005, **XIV**.
35. Ludwig TE, Levenstein ME, Jones JM, Berggren WT, Mitchen ER, Frane JL, Crandall LJ, Daigh CA, Conard KR, Piekarczyk MS, Llanas RA, Thomson JA: **Derivation of human embryonic stem cells in defined conditions.** *Nat Biotechnol* 2006, **24**:185-187.
36. American College of Cardiology: **Press release.** 2006.
37. Wadman M: **Medical schools swap pigs for plastic.** *Nature* 2008, **453**:140-141.
38. Noebels JL: **Single gene models of epilepsy** In *Jasper's Basic Mechanisms of the Epilepsies Volume* 79. 3rd edition. Edited by Delgado-Escueta AV, Wilson WA, Olsen RW, Porter RJ; 1999: 227-238: *Advances in Neurology*].
39. Jankovic J, Noebels JL: **Genetic mouse models of essential tremor: are they essential?** *J Clin Invest* 2005, **115**:584-586.
40. Gibbs RA, Rogers J, Katze MG, Bumgarner R, Weinstock GM, Mardis ER, Remington KA, Strausberg RL, Venter JC, Wilson RK, Batzer MA, Bustamante CD, Eichler EE, Hahn MW, Hardison RC, Makova KD, Miller W, Milosavljevic

A, Palermo RE, Siepel A, Sikela JM, Attaway T, Bell S, Bernard KE, Buhay CJ, Chandrabose MN, Dao M, Davis C, Delehaunty KD, Ding Y, Dinh HH, Dugan-Rocha S, Fulton LA, Gabisi RA, Garner TT, Godfrey J, Hawes AC, Hernandez J, Hines S, Holder M, Hume J, Jhangiani SN, Joshi V, Khan ZM, Kirkness EF, Cree A, Fowler RG, Lee S, Lewis LR, Li Z, Liu YS, Moore SM, Muzny D, Nazareth LV, Ngo DN, Okwuonu GO, Pai G, Parker D, Paul HA, Pfannkoch C, Pohl CS, Rogers YH, Ruiz SJ, Sabo A, Santibanez J, Schneider BW, Smith SM, Sodergren E, Svatek AF, Utterback TR, Vattathil S, Warren W, White CS, Chinwalla AT, Feng Y, Halpern AL, Hillier LW, Huang X, Minx P, Nelson JO, Pepin KH, Qin X, Sutton GG, Venter E, Walenz BP, Wallis JW, Worley KC, Yang SP, Jones SM, Marra MA, Rocchi M, Schein JE, Baertsch R, Clarke L, Csuros M, Glasscock J, Harris RA, Havlak P, Jackson AR, Jiang H, Liu Y, Messina DN, Shen Y, Song HX, Wylie T, Zhang L, Birney E, Han K, Konkel MK, Lee J, Smit AF, Ullmer B, Wang H, Xing J, Burhans R, Cheng Z, Karro JE, Ma J, Raney B, She X, Cox MJ, Demuth JP, Dumas LJ, Han SG, Hopkins J, Karimpour-Fard A, Kim YH, Pollack JR, Vinar T, Addo-Quaye C, Degenhardt J, Denby A, Hubisz MJ, Indap A, Kosiol C, Lahn BT, Lawson HA, Marklein A, Nielsen R, Vallender EJ, Clark AG, Ferguson B, Hernandez RD, Hirani K, Kehrer-Sawatzki H, Kolb J, Patil S, Pu LL, Ren Y, Smith DG, Wheeler DA, Schenck I, Ball EV, Chen R, Cooper DN, Giardine B, Hsu F, Kent WJ, Lesk A, Nelson DL, O'Brien W E, Prufer K, Stenson PD, Wallace JC, Ke H, Liu XM, Wang P, Xiang AP, Yang F, Barber GP, Haussler D, Karolchik D, Kern AD, Kuhn RM, Smith KE, Zwieg AS: **Evolutionary and biomedical insights from the rhesus macaque genome.** *Science* 2007, **316:**222-234.

41. Threadgill DW, Dlugosz AA, Hansen LA, Tennenbaum T, Lichti U, Yee D, LaMantia C, Mourton T, Herrup K, Harris RC, et al.: **Targeted disruption of mouse EGF receptor: effect of genetic background on mutant phenotype.** *Science* 1995, **269:**230-234.
42. Pearson H: **Surviving a knockout blow.** *Nature* 2002, **415:**8-9.
43. Snyder L: **Is Evidence Historical?** In *Philosophy of Science.* Edited by Curd M, Cover J: Norton; 1998: 460-480
44. Salmon W: **Rational Prediction.** In *Philosophy of Science.* Edited by Curd M, Cover J: Norton; 1998: 433-444
45. Suter K: **What can be learned from case studies? The company approach.** In *Animal Toxicity Studies: Their Relevance for Man.* Edited by Lumley C, Walker S. Lancaster: Quay; 1990: 71-78
46. Heywood R: **Clinical Toxicity--Could it have been predicted? Post-marketing experience.** In *Animal Toxicity Studies: Their Relevance for Man.* Edited by CE Lumley, Walker S. Lancaster: Quay; 1990: 57-67
47. Spriet-Pourra. C, Auriche. M: **Drug Withdrawal from Sale.** 2nd edition. New York; 1994.
48. Fletcher AP: **Drug safety tests and subsequent clinical experience.** *J R Soc Med* 1978, **71:**693-696.
49. Lumley C: **Clinical toxicity: could it have been predicted? Premarketing experience.** In *Animal Toxicity Studies: Their Relevance for Man.* Edited by Lumley C, Walker S: Quay; 1990: 49-56
50. Olson H, Betton G, Robinson D, Thomas K, Monro A, Kolaja G, Lilly P, Sanders J, Sipes G, Bracken W, Dorato M, Van Deun K, Smith P, Berger B, Heller A: **Concordance of the toxicity of pharmaceuticals in humans and in animals.** *Regul Toxicol Pharmacol* 2000, **32:**56-67.

51. Wilbourn J, Haroun L, Heseltine E, Kaldor J, Partensky C, Vainio H: **Response of experimental animals to human carcinogens: an analysis based upon the IARC Monographs programme.** *Carcinogenesis* 1986, **7**:1853-1863.
52. Rall DP: **Laboratory animal tests and human cancer.** *Drug Metab Rev* 2000, **32**:119-128.
53. Tomatis L, Aitio A, Wilbourn J, Shuker L: **Human carcinogens so far identified.** *Jpn J Cancer Res* 1989, **80**:795-807.
54. Schardein J: *Drugs as Teratogens.* CRC Press; 1976.
55. Schardein J: *Chemically Induced Birth Defects.* Marcel Dekker; 1985.
56. Manson J, Wise D: **Teratogens.** In *Casarett and Doull's Toxicology.* 4th edition; 1993: 228
57. Staples RE, Holtkamp DE: **Effects of Parental Thalidomide Treatment on Gestation and Fetal Development.** *Exp Mol Pathol* 1963, **26**:81-106.
58. Runner MN: **Comparative pharmacology in relation to teratogenesis.** *Fed Proc* 1967, **26**:1131-1136.
59. Keller SJ, Smith MK: **Animal virus screens for potential teratogens. I. Poxvirus morphogenesis.** *Teratog Carcinog Mutagen* 1982, **2**:361-374.
60. Lin JH: **Species similarities and differences in pharmacokinetics.** *Drug Metab Dispos* 1995, **23**:1008-1021.
61. Eason CT, Bonner FW, Parke DV: **The importance of pharmacokinetic and receptor studies in drug safety evaluation.** *Regul Toxicol Pharmacol* 1990, **11**:288-307.
62. Coulston F: **Final Discussion.** In *Human Epidemiology and Animal Laboratory Correlations in Chemical Carcinogenesis.* Edited by Coulston F, Shubick P: Ablex; 1980: 407
63. Staples RE: **Predictiveness and limitations of test methods in teratology: overview.** *Environ Health Perspect* 1976, **18**:95-96.
64. Shepard TH: *Catalog of Teratogenic Agents.* Johns Hopkins University Press; 1995.
65. Schardein JL: *Chemically Induced Birth Defects.* 2nd edn: Marcel Dekker; 1992.
66. [http://www.amprogress.org/site/c.jrLUK0PDLoF/b.933657/k.923A/ANI-MAL_RESEARCH.htm]
67. Marshall E: **Gene therapy on trial.** *Science* 2000, **288**:951-957.
68. Fuchs VR, Sox HC, Jr.: **Physicians' views of the relative importance of thirty medical innovations.** *Health Aff (Millwood)* 2001, **20**:30-42.
69. Fenster A: **A TRENDS Guide to Imaging Technologies.** *Trends in Biotechnology* 2002, **20**:S1-S2.
70. Paul JR: *A History of Poliomyelitis.* New Haven: Yale University Press; 1971.
71. Sabin A: **Testimony before the subcommittee on Hospitals and Health Care, Committee on Veterans Affair's, House of Representatives, April 26, 1984 serial no. 98-48.**
72. American Medical Association: *White Paper on Animal Research.* American Medical Association; 1992.
73. Roberts F: **Insulin.** *BMJ* 1922:1193-1194.
74. Weisse AB: **The long pause. The discovery and rediscovery of penicillin.** *Hosp Pract (Off Ed)* 1991, **26**:93-96, 101-104, 107 passim.
75. Steffee CH: **Alexander Fleming and penicillin. The chance of a lifetime?** *N C Med J* 1992, **53**:308-310.

76. Hare R: **Hare R. Uncataloged archives. Wellcome Institute for the History of Medicine. Letter, December 6, 1955 from Dolman to Hare, Letter December 29, 1955 from Rogers to Hare, Letter June 12, 1955 from Craddock to Hare..**
77. Swan H: **Medicine in Sheffield.** *Q J Med* 1992, **296:**1041-1049.
78. Wainwright M: **The mystery of the plate: Fleming's discovery and contribution to the early development of penicillin.** *J Med Biogr* 1993, **1:**59-65.
79. Diggins FW: **The true history of the discovery of penicillin, with refutation of the misinformation in the literature.** *Br J Biomed Sci* 1999, **56:**83-93.
80. Henderson JW: **The yellow brick road to penicillin: a story of serendipity.** *Mayo Clin Proc* 1997, **72:**683-687.
81. Florey H, Chain E, Healey N, et al.: *Antibiotics : A Survey of Penicillin, Streptomycin, and Other Antimicrobial Substances from Fungi, Actinomycetes, Bacteria and Plant. Vol II.* Oxford: Oxford University Press; 1949.
82. Florey H: **The advance of chemotherapy by animal experiment.** *Conquest* 1953, **41:**12.
83. Harare DM, Rake C, McKee C, M., MacPhillamy HB: **The toxicity of penicillin as prepared for clinical use.** *Am J M Sc* 1943, **206:**642-652.
84. Schneierson SS, Perlman E: **Toxicity of penicillin for the Syrian hamster.** *Proc Soc Exp Biol Med* 1956, **91:**229-230.
85. Kamb A: **What's wrong with our cancer models?** *Nat Rev Drug Discov* 2005, **4:**161-165.
86. Clemmensen J, Hjalgrim-Jensen S: **On the absence of carcinogenicity to man of phenobarbital.** In *Human Epidemiology and Animal Laboratory Correlations in Chemical Carcinogenesis.* Edited by Coulston F, Shubick S: Alex Pub.; 1980: 251-265
87. Northrup E: *Science looks at smoking: A new inquiry into the effects of smoking on your health.* . New York: Coward-McCann; 1957.
88. Utidjian M: In *Perspectives in Basic and Applied Toxicology.* Edited by Ballantyne B: Butterworth-Heinemann; 1988
89. Janofsky M: **On Cigarettes, Health and Lawyers.** In *New York Times.* New York; 1993.
90. [http: www.publications.parliament.uk/pa/ld200102/idselect/ldanimal/999/2042302.htm.]
91. Shapiro K: **Animal Model Research. The Apples and Oranges Quandry.** *ATLA* 2004, **32:**405-409.
92. Shanks N, Pyles RA: **Evolution and medicine: the long reach of "Dr. Darwin".** *Philos Ethics Humanit Med* 2007, **2:**4.
93. Shubin N, Tabin C, Carroll S: **Fossils, genes and the evolution of animal limbs.** *Nature* 1997, **388:**639-648.
94. *Animal People*, December edition. pp. 17; 2004:17.
95. Editorial: **Devil in the details.** *Nat Med* 2008, **14:**105-106.
96. Davis MM: **A prescription for human immunology.** *Immunity* 2008, **29:**835-838.
97. Hughes B: **Industry concern over EU hepatotoxicity guidance.** *Nat Rev Drug Discov* 2008, **7:**719-719.
98. Schnabel J: **Neuroscience: Standard model.** *Nature* 2008, **454:**682-685.
99. Begley S: **A New Reason to Frown. Does Botox get into the brain? Troubling research contradicts earlier findings about the treatment.** In *Newsweek*; 2008.
100. Butcher EC: **Can cell systems biology rescue drug discovery?** *Nat Rev Drug Discov* 2005, **4:**461-467.

101. Curry SH: **Why have so many drugs with stellar results in laboratory stroke models failed in clinical trials? A theory based on allometric relationships.** *Ann N Y Acad Sci* 2003, **993:**69-74; discussion 79-81.
102. Committee on Toxicity Testing and Assessment of Environmental Agents NRC: *Toxicity Testing for Assessment Agents: Interim Report.* National Academies Press; 2006.
103. Lindl T, Voelkel M, Kolar R: **[Animal experiments in biomedical research. An evaluation of the clinical relevance of approved animal experimental projects].** *ALTEX* 2005, **22:**143-151.
104. Höerig H, Pullman W: **From bench to clinic and back: Perspective on the 1st IQPC Translational Research conference.** *J Transl Med* 2004, **2:**44.
105. Brady CA: **Of Mice and Men: the potential of high-resolution human immune cell assays to aid the pre-clinical to clinical transition of drug development projects.** *Drug Discovery World* 2008/9:74-78.
106. Mohnmaney T: **Marshall's Hunch.** In *New Yorker*, vol. 69. pp. 64-72; 1993:64-72.
107. Pigliucci M: *Denying Evolution.* Sinauer; 2002.
108. Gad S: **Preface.** In *Animal Models in Toxicology.* Edited by Gad S: CRC Press; 2007: 1-18
109. Hau J: **Animal Models.** In *Handbook of Laboratory Animal Science Animal Models. Volume* II. 2nd edition. Edited by Hau J, van Hoosier Jr GK: CRC Press; 2003: 2-8
110. Anonymous: **Of Mice...and Humans.** *Drug Discovery and Development* 2008, **11:**16-20.
111. Fomchenko EI, Holland EC: **Mouse models of brain tumors and their applications in preclinical trials.** *Clin Cancer Res* 2006, **12:**5288-5297.
112. Rowan A: **Book Review. Brute Science.** *Animal Welfare* 1997, **6:**378-381.
113. Smith C: **The Naked Scientist Podcast.** In *Fusion: The Power of the Sun.* pp. begin 1:45 October, 19 2008:begin 1:45
114. **FDA Issues Advice to Make Earliest Stages Of Clinical Drug Development More Efficient** [http://www.fda.gov/bbs/topics/news/2006/NEW01296.html]
115. Moynihan R, Bero L, Ross-Degnan D, Henry D, Lee K, Watkins J, Mah C, Soumerai SB: **Coverage by the news media of the benefits and risks of medications.** *N Engl J Med* 2000, **342:**1645-1650.
116. **Drug Cheerleaders.** *New Scientist* 2000:19.
117. Sloan RP, Bagiella E, Powell T: **Religion, spirituality, and medicine.** *Lancet* 1999, **353:**664-667.
118. **Correction: Looking Back on the Millennium in Medicine.** *N Engl J Med* 2000, **342:**988.
119. Shermer M: *Scientific American* 2001, **Nov and Dec.**
120. Conn P, Parker J: *The Animal Research War.* Palgrave; 2008.
121. Botting JH, Morrison AR: **Animal research is vital to medicine.** *Sci Am* 1997, **276:**83-85.
122. Trull F: *Animal Models: Assessing the Scope of Their Use in Biomedical Research.* Charles River, MA: Charles River; 1987.
123. Foundation for Biomedical Research: **Animal Research Fact vs. Myth.**
124. Foundation for Biomedical Research: Foundation for Biomedical Research; 1992.
125. Xi S: **Sigma Xi Statements of the Use of Animals in Research.** *American Scientist*, **80:**73-76.

126. Garattini S and van Bekkum: *The Importance of Animal Experiments for Safety and Biomedical Research.* Dordrecht Kluwer Academic Publishers; 1990.
127. **Critical Thought** [http://www.criticalthinking.org/aboutCT/define_critical_thinking.cfm]
128. Conn PM, Parker JV: **Winners and Losers in The Animal-Research War.** *American Scientist* 2008, **May-June.**
129. Shanks N, Greek R, Nobis N, Greek J: **Animals and Medicine: Do Animal Experiments Predict Human Response?** *Skeptic* 2007, **13**:44-51.
130. Shanks N, Greek R: *Animal Models in Light of Evolution.* Brown Walker; 2009.
131. [www.curedisease.com]

Index of Questions

About the Authors

Ray Greek received his MD from the University of Alabama at Birmingham in 1985 and completed a residency in anesthesiology in 1989 at the University of Wisconsin. He has taught at the medical schools of the University of Wisconsin and Thomas Jefferson University in Philadelphia. He has performed research with animals and humans. He is the president and co-founder of Americans For Medical Advancement (AFMA). Greek enjoys scuba diving, swimming with pinnipeds and cetaceans, interacting with African wildlife and photographing same. While he enjoys those activities, what he actually does with his time is take care of all the animals Jean brings home.

Niall Shanks is the Curtis D. Gridley Distinguished Professor of History and Philosophy of Science, Wichita State University. He received a PhD in Philosophy in 1987 from the University of Alberta. He is the vice-president of AFMA, and was president of the Southwestern and Rocky Mountain Division of the American Association for the Advancement of Science for 2008-2009. His publications include *God, the Devil and Darwin* (Oxford University Press 2004). Shanks is currently learning to play the bagpipes. He shares his domicile, his paycheck, and his life with his dogs Brutus, the Lummocks, and Gnasher. All four are fond of meat. Only Shanks enjoys the bagpipes!

AFMA is a not-for-profit organization that promotes biomedical research and the practice of medicine based on critical thinking and our current understanding of evolutionary and developmental biology, complex systems, genomics, and science in general. AFMA's position on the use of animals in science as explained in this book is that animals can be successfully used in numerous scientific endeavors but not as predictive models for humans.